D1645168

26 0128829 8

Spectrum Wars

The Policy and Technology Debate

Jennifer A. Manner

Artech House
Boston • London
www.artechhouse.com

For a listing of recent titles in the *Artech House Telecommunications Library*, turn to the back of this book.

Spectrum Wars

The Policy and Technology Debate

Library of Congress Cataloging-in-Publication Data

A catalog record for this book is available from the Library of Congress.

British Library Cataloguing in Publication Data

A catalog record of this book is available from the British Library.

Cover design by Gary Ragaglia

685 Canton Street
Norwood, MA 02062

International Standard Book Number: 1-58053-483-x
A Library of Congress Catalog Card Number is available from the British Library.
10 9 8 7 6 5 4 3 2 1

In loving memory of my father and my hero, Maurice R. Manner

Contents

Acknowledgments xiii

Introduction xv

1 An Overview 1

Introduction 2

An overview 4

International overview 5

Domestic issues 6

Spectrum terminology 8

A brief background of spectrum management 11

Companies, governments, and other interests 16

Incumbent and planned/new users 17

The various interests 17

An example of the scientific community: radio astronomers 20

Public safety uses 21

Two unique battles for the spectrum resource for new services 21
NGSO FSS: taking the world by storm 21
Third generation mobile service and the FS: a compromise 25
Conclusion 29
Endnotes 29

2 Spectrum Primer 33
Overview of technical characteristics of the radiocommunications spectrum resource 34
Spectrum scarcity and harmful interference 37
The allocation scheme 40
Key technical considerations when evaluating spectrum use 42
Other considerations 46
Endnotes 47

3 Radio Communications Spectrum and Telecommunications Players 49
Wireless versus wireline network solutions 49
The key participants 54
Government use 55
Telecommunications service providers and broadcasters 57
Telecommunications equipment manufacturers 60
Consumers 62
Factors impacting the use of the spectrum resource 63
The government regulator and the accompanying regulatory regime . . 64
Market demand for the service 66
Amount of spectrum available for the same or similar use 67
The costs of obtaining access to the spectrum and the impact on the business case 68
The availability of terrestrial wireline infrastructure 68
Conclusion 69
Endnotes 69

4 The Regulatory Regime Governing Spectrum 71

Why is the radiocommunications spectrum resource regulated?. . . . 71
The goals of spectrum regulation. 74
The governing regulatory bodies 76
The international regulatory process. 77
An overview of the ITU and the Radiocommunications Sector 78
Overview of the ITU R Sector 79
The impact of regional organizations on spectrum regulation. 81
The international spectrum allocation process 83
Technical issues 85
An overview of the WRC process 87
Conclusion 89
Endnotes. 90

5 Domestic Regulation of Spectrum, Part I: International Representation. 93

Overview of the domestic regulation of the radiocommunications spectrum resource 93
Participation in the international arena 94
Domestic participants in the international process 95
The domestic preparatory process for international meetings. 98
Endnotes 103

6 Domestic Regulation of Spectrum, Part II: Allocation, Assignment, and Use 105

Overview 105
Important cornerstones of domestic regulation 107
Impact of the WTO Agreement 108
The domestic allocation of frequency bands to individual services . . 110
The identification or designation of radiocommunications spectrum to specified uses 112
Changes to rules 112
Flexible use 113
Relocation of existing users 114

The assignment and authorization of spectrum to specific users . . . 116
Overview of assignment processes 118
The implementation and enforcement of technical and operating rules 127
The regulation of secondary markets 128
Conclusion 129
Endnotes 130

7 Solutions to Harmful Interference 131

Overview 131
Spectrum conflict: the potential for harmful interference 133
Minimizing the potential for conflicts 135
An anomaly: unlicensed spectrum usage 138
Regulatory mechanisms to adopt rules governing cofrequency sharing, frequency band segmentation, and relocation 140
Cofrequency sharing and frequency-band segmentation: an overview 143
Frequency-band segmentation 146
Relocation of existing users 147
Overview 147
The need for comparable spectrum and reasonable compensation for relocated uses 148
Endnotes 150

8 Secondary Markets for Spectrum 151

The increasing use of secondary markets 151
Advantages and disadvantages to the use of secondary spectrum markets 152
Types of secondary spectrum market regimes 155
New Zealand: an overview 156
Creating a regime governing secondary markets for spectrum 158
Conclusion 161
Endnotes 161

9 Impact of the Telecommunications Financial Crisis . 163
Key reasons for the telecommunications financial meltdown 165
Impact of the telecommunications meltdown 166
The rebuilding of an industry. 168
Can wireless service providers fare better? 170
Endnotes 172

Appendix A: List of Web Addresses 173

About the Author 177

Index. 179

Acknowledgments

I would like to thank several people for their input into and their assistance with this book. First, I would like to thank Donna Bethea of PanAmSat Corporation. Donna has spent countless hours debating the issues addressed herein with me, as I have tried to work through some of the tough policy questions concerning access to the spectrum. In addition, Donna provided technical insight into many of the key issues addressed in this book. Donna and I have lived through numerous spectrum battles together, and it is often because of her insight that they have turned out successfully.

Second, I would like to thank Barbara Englehart, my colleague at WorldCom, Inc., for her insights on the financial meltdown of the telecommunications market.

It is with a very heavy heart that I write my next acknowledgment, to my parents, Joyce and Maurice Manner. This past year my father passed away after a short fight with a terrible disease, endocarditis. Prior to his death, both my father and mother provided editorial assistance on this book and my first book. They read each chapter in detail and provided valuable comments. My mother bravely carried on this effort singularly after my father's tragic death. In addition, my mother provided me with much-needed research support on some key issues. Without the support and love of my parents, I would have not have been able to succeed in any of my achievements.

Finally, I would like to thank my husband, Eric Glasgow, for always being there for me, whatever my latest endeavor. Eric spent countless hours helping me think and rethink many of the issues in this book, often debating various points of view. He is also, most importantly, my inspiration.

Introduction

In early 2002, when the idea for this book was born and initial drafting began, the outlook for the telecommunications industry, including the wireless world, was beginning to look bleak, especially in comparison to the growth days of the late 1990s. Certain wireless telecommunications companies, such as Winstar and Metricom, began to face financial trouble, but the truly wide-ranging problems that began to impact the entire telecommunications market in mid-2002 were not yet evident. Few, if any, experts foresaw the impending bankruptcy of the telecommunications giant, WorldCom, or the dissolution of other major telecommunications companies, such as KPN Qwest and Teleglobe.

There are many potential reasons for this decline in the telecommunications industry, including the blame of its own players for overzealousness in expansion and profit making, as will be discussed in Chapter 9 of this book. However, whatever the reasons, what is clear is that the telecommunications industry, although caught in a retrenchment, will rebuild itself, albeit in a different manner and will reemerge as a vibrant sector. This includes the wireless telecommunications industry that is in the midst of reshaping and retrenching itself to reflect the realities of doing business in the twenty-first century.

As the telecommunications industry is rebuilt, there are certain "givens" that can be relied upon:

- The industry will survive, especially because of the reliance of the global population on the critical services provided by the industry and the need for the public, businesses, and the government to communicate.
- There will be a retrenchment period in the telecommunications industry, at least for the next few years. This will mean, at least for the foreseeable future, a reduction in the number of large, medium, and small players in the industry.
- Resources for expansion by the telecommunications industry into new geographic areas or product lines will be scarcer. This will mean telecommunications companies will have a harder standard to meet to justify such expansion within the finance community.
- Funding in general for the telecommunications industry will be harder to obtain, and it is likely that the financial industry will impose stricter scrutiny on such investments.
- Because of the stricter financial scrutiny, telecommunications services that are generally less capital intensive with a large customer demand, such as wireless services, will become increasingly attractive to deploy.

This last reason is key to the belief that the wireless industry and services that utilize wireless services will continue to grow in profitability, even with the telecommunications and general economic downturn. However, it is also likely that these entrants will need to carefully structure entry and operation in a manner that makes financial sense. This will include formulating cost-effective, efficient plans to achieve their desired goals, including obtaining access and approval to operate in relevant frequency bands of the radiocommunications spectrum resource. Further, governments may need to restructure their regulatory regimes in a way that recognizes these market realities.

With this change in the direction of the telecommunications industry, in order to be well positioned to successfully participate in the battles over access to the radiocommunications spectrum resource, it is critical to understand the regime governing this resource. This book provides such insight. Specifically, this book provides an understanding of the radiocommunications spectrum resource itself, an overview of the international and domestic regulatory processes governing spectrum, and a detailed discussion of the different ways in which portions of the radiocommunications spectrum are allocated, assigned, and utilized. In addition, this book explores many of the difficult policy questions that are being faced in the spectrum arena, including issues concerning the use of secondary spectrum markets. Finally, this book concludes with a discussion of the financial meltdown of the telecommunications industry and its impact on the battles over obtaining access to the radiocommunications spectrum.

1

An Overview

[W]ireless is exciting because it's at the cutting edge of innovation. I think it's at the cutting edge of competitive principles.

—FCC Chairman Michael Powell [1]

Increasingly innovative wireless telecommunications services are among the most exciting technologies to be introduced into the marketplace in the late twentieth and early twenty-first centuries. In order for wireless services to operate, they must have access to a discrete portion of the radiocommunications spectrum resource. Radiocommunications can be defined as a radio emission or receipt of a radio emission for the purposes of communicating. More specifically, the radiocommunications spectrum refers to the range of frequencies of electromagnetic radiation within which radiocommunications can occur. For example, mobile telephones operate in a discrete portion of the radiocommunications spectrum, generally below the 3-GHz range. The focus of this book is on the regime governing the radiocommunications spectrum resource and the fierce efforts of wireless service providers and other interested parties to gain access to this valuable resource.

Introduction

Radiocommunications spectrum is an intangible commodity that continues to grow in its importance as a critical component to the successful deployment of telecommunications technologies and services. Its importance is built around many factors. One key factor is that the use of technology can eliminate the need, in part or in whole, for the wireline infrastructure that has traditionally bound communications to the static and often costly wired network [2]. By utilizing wireless networks, mobile and fixed communications are increasingly readily available, often on a more economical basis and in more remote locations. Further, radiocommunications spectrum is the only transmission mechanism for the successful deployment and operation of many of the most advanced telecommunications technologies and services, such as third generation (3G) mobile service, as well as a mainstay of existing spectrum uses, including public safety uses, ship-to-shore communications, and aeronautical communications. In other cases, wireless services serve as adjuncts to existing wireline networks.

Access to certain discrete portions of the radiocommunications spectrum is essential for new technological advances to reach fruition in the wireless world. Without this access, it is possible that many of the promised and most fantastic technologies will not be developed. In addition, continued access to the spectrum resource is essential for existing technologies to continue their operations. Adequate access by new or existing technologies directly impacts the bottom line of service providers who are reliant on the use of the radiocommunications spectrum resource. Accordingly, the stakes are often quite high as parties seek to obtain the right to utilize the spectrum.

Another fundamental aspect is that spectrum can be heavily or lightly utilized. This means that regulators may by design or accident create a regime that results in more congestion in certain frequency bands. For example, if a regulator does not charge a monetary fee for spectrum and does not impose efficiency requirements on operating systems, the designer of the relevant systems may choose not to utilize efficient technology (e.g., such that spectrum reuse is widely available) to design its systems. Such a system may result in the development of unnecessary spectrum congestion.

Because of the real or perceived scarcity of the radiocommunications spectrum resource, as well as the value public- and private-sector telecommunications operators and consumers place on its use, the radiocommunications spectrum is a commodity over which costly and at times vicious domestic and international battles have been and will continue to be fought. Two notable battles in recent history involved the search for spectrum for use by the *nongeostationary orbit fixed satellite service* (NGSO FSS) in the Ka band and the identification of spectrum for use by 3G mobile services in the 2-GHz band. Both of these endeavors involved the expenditure of large amounts of financial and political resources on the part of the spectrum advocates, their supporters, and their opponents, among others. This is in contrast to most spectrum wars, where substantially fewer resources are available to the advocates to expend. However, they do exemplify the extents to which spectrum advocates will go in order to achieve their goals. In both cases, the spectrum advocates and their allies expanded large sums of money to enlist the necessary support to ensure that their proposed telecommunications systems were able to gain access to the specific frequency band they sought, with accompanying technical rules to ensure their optimal operation.

Oftentimes obtaining an authorization for deploying a wireless system is a battle in and of itself when there are competing uses by applicants for the use of the same spectrum. In order to resolve this issue, governments may rely on any of several different mechanisms, including lotteries and competitive bidding procedures (also known as auctions). Since the mid-1990s, competitive bidding procedures have been the solution of choice in these situations by most governments, as they generally lead to what is considered a nonbiased result with a large financial incentive for the government. Auctions have often resulted in huge sums of money being paid for use of spectrum, such as for 3G services. For example, the British Telecommunications bid of £4,030,100,000 for an authorization to provide 3G services in the United Kingdom astounded many experts [2] (see Table 1.1).

Having a clear understanding of how radiocommunications spectrum is allocated, assigned, and awarded for use by service providers, network operators, and other interests is an important tool in order to success fully maneuver in today's increasingly wireless world. This book addresses the key areas associated with spectrum allocation, assignment, and use—including the myriad economic, technical, regulatory, and political issues—from both a domestic and an international perspective. Many of these issues will be addressed utilizing case studies, including the cutting-edge issues associated with the allocation and use of radiocommunications spectrum for advanced wireless services, including 3G services.

Table 1.1
United Kingdom 3G Auction Results (for Five Authorizations)

Company	Successful Bid
TIW	£4,384,700,000
Vodafone	£5,964,000,000
British Telecommunications	£4,030,100,000
One2One	£4,003,600,000
Orange	£4,095,000,000

Source: [3].

In addition, this book will also explore the impact of the recent telecommunications downfall that has been occurring since the beginning of the twenty-first century and what it means for the wireless world, including how it might shape future spectrum battles. Further, it will also look at some of the more novel ideas proposed to redesign the spectrum regime through the use of recently proposed solutions, such as the creation of secondary spectrum markets and the use of *gray* spectrum.

The purpose of this chapter is to provide a brief introduction and overview of the areas that will be focused on in this book, as well as some of the key terms and concepts that are integral to understanding and structuring strategies to successfully participate in a spectrum war. This chapter should provide a firm basis for delving into the subsequent chapters of this book for even those unfamiliar with the battles associated with the radiocommunications spectrum resource.

An overview

Radiocommunications spectrum is one the most valuable scarce resources in the world, and it is a requirement for many of the crucial telecommunications services of the twenty-first century. Like many of the world's natural resources, radiocommunications spectrum is extremely limited in its availability for exploitation. However, there is a critical difference between radiocommunications spectrum scarcity and the scarcity commonly associated with commodity resources such as oil or diamonds. In the case of radiocommunications spectrum, as with airspace, multiple or increased usage can be maximized through sound resource management, such as putting in place rational allocation schemes and accompanying technical requirements, or requiring that more efficient, albeit often expensive, technologies be utilized by spectrum users. In addition, improving technologies that allow access to the more remote sections of the spectrum or shared use of a frequency band or bands enables the spectrum resource to have further expanded capabilities and accommodate increasingly larger amounts of users. Accordingly, it is imperative then that sound resource management is utilized by government regulators and relevant international organizations to ensure that the radiocommunications spectrum is utilized efficiently. Because the private sector and other users of the spectrum resource may not have the same goals, the government may need to impose regulations that ensure efficient usage and access to the spectrum. Often, such regulation must be accompanied by an appropriate enforcement mechanism or it will be unsuccessful in achieving its goals.

International overview

The radiocommunications spectrum resource does not recognize artificial country boundaries. Accordingly, many key spectrum issues are resolved internationally as well as domestically. Because of the scarcity issues associated with the radiocommunications spectrum, the overarching goal by governments in managing the electromagnetic spectrum should be to maximize its use to both the country and the world's population. This should be done in a manner that is both consistent with international treaty obligations and that ensures the availability of the radiocommunications spectrum to mankind through efficient use. In the international arena, this goal has been sought to be fulfilled through the *International Telecommunication Union* (ITU) spectrum allocation and regulatory process.

As discussed subsequently in greater depth, the ITU is an arm of the United Nations. The ITU Constitution stipulates that the ITU is a treaty-making body that, among other functions, allocates use of the radiocommunications spectrum resource among various radio services. Allocation and use issues among different radio services often result in fierce battles among the ITU member states. These battles are almost always settled by international compromise among the interested parties through technical and political solutions, and they ultimately form the treaty obligations governing the radiocommunications spectrum resource.

Once settled, member countries are obligated—because of the treaty nature of these regulations—to abide by the agreed allocation scheme and allow nonconforming uses only where they will not cause harmful interference to the in-use services of another country [4]. However, it should be noted that although there is a treaty obligation to comply with the adopted allocations and accompanying regulations, there is no formal enforcement mechanism if a country does not comply. However, compliance generally occurs because failure to abide by the agreed-upon allocations and the accompanying technical rules could result in harmful interference among operating systems in multiple countries, leading to the inability of these systems to operate free from harmful interference. Therefore, each country has a keen self-interest in compliance or chaos will likely ensue. In addition, there is often vast political pressure existing between member states to encourage compliance in general.

A good example of the reason for such compliance is demonstrated by examining regions such as Western Europe, where country borders are so close together that failure to adhere to ITU regulations would likely result in harmful interference between multiple purposes. Similarly, other countries, such as Japan, where the magnitude of such cross-border concerns is less, still have the incentive to comply because of the likely desire to facilitate increased trade in telecommunications equipment across many countries. Accordingly, countries are willing to expend significant political resources to ensure that allocations and accompanying technical rules are generally followed by ITU member states. These reasons also help to explain why the stakes in spectrum battles are so high.

Domestic issues

While international allocations and accompanying technical rules are subject to a treaty obligation, the lack of an express enforcement mechanism contained within the ITU Convention means that they obtain their real teeth from domestic compliance and implementation. Domestically, however, because of the jurisdiction of national authorities to award the use of the spectrum resource to specific purposes, varying forms of allocation, assignment, and authorization have been relied upon. These are discussed briefly next and will be subject of much greater discussion throughout this book.

In the more liberalized telecommunications markets, governments will often hold public proceedings to determine the best use of the spectrum domestically, generally in compliance with their international obligations. Examples of countries that hold such proceedings include Germany, the United Kingdom, and Brazil. In some cases, these domestic proceedings may be quite contentious, as new entrants and incumbent users of the same spectrum often have competing views on what use is the most appropriate for the spectrum. The outcomes, as in the international arena, are as likely to be reliant on technical, economic, and political considerations.

The first step the domestic regulator must make is in determining the overall use of a particular frequency band. In some instances, such as in telecommunications markets that are largely closed to competition, such as in China or Afghanistan, governments will often make unilateral decisions concerning use of the spectrum, sometimes with little regard for current and planned uses of the frequency band. Such a decision-making process often leads to inefficient use of the radiocommunications spectrum resource or results in the inability of new technologies to operate successfully in the allocated or assigned frequency bands, as these decisions may be primarily politically based or based on other nontransparent rationales. However, as telecommunications markets continue to open to competition and become liberalized, more and more countries are using public proceedings to determine the appropriate domestic allocation and assignment of the spectrum resource. The use of public proceedings and other procedures discussed subsequently help to ensure that decisions concerning the use of discrete portions of the radiocommunications resource are made in a rational manner.

The authorization process that is utilized may also impact the ability of spectrum advocates to gain access to desired portions of the radiocommunications resource. In noncompetitive telecommunications marketplaces, the government generally makes unilateral decisions as to who can operate in a specific frequency band. Often such decisions occur in a nontransparent manner.

In partially or fully competitive markets, governments may rely on several different methods when issuing service authorizations. First, when there are no competing uses, and often when there is not much demand for a discrete portion of the radiocommunications spectrum, a straight authorization process is generally utilized. However, if there are competing uses or applications, regulators must develop and implement an authorization process to choose one use or applicant over another. Some regulators believe that the best means of achieving efficient use of the spectrum is to rely principally on market forces, such as that which occurs through the use of competitive bidding or auctions processes. This approach has been the preferred approach of many regulators from more liberalized markets in authorizing mobile telephony service providers and more recently 3G service providers, including France, Japan, and Singapore.

Other regulators, however, believe that this approach is mistaken and will lead to inefficient use of the spectrum resource. These regulators may turn to more traditional mechanisms of awarding authorizations, such as comparative hearings, lotteries, the creation of frequency band segmentation schemes, or a novel authorization framework. As will be discussed in later chapters, if the recent financial fallout from spectrum auctions continues, more use of nonmarket-based or other new, innovative authorization schemes are likely to be instituted by regulators that will ensure that the radiocommunications spectrum resource does not lay fallow.

This book will continue to build on this examination of how spectrum has been allocated, assigned, and put into use. Subsequent portions of this book also address how spectrum can be managed to ensure its efficient use, how to acquire access to it, and how its use can be maximized. As will be seen, many different approaches have been attempted in both the domestic and international arenas. As demonstrated in the accompanying case studies, some of these efforts have been more successful than others, and other approaches may be available in the future.

Spectrum terminology

Before delving in more detail into the issues surrounding spectrum wars, it is important to understand the regulatory terminology associated with the use of the radiocommunications spectrum. Essentially, many different categories, as depicted in Table 1.2, are widely recognized as ways to determine the use of spectrum.

TABLE 1.2
Key Spectrum Terms

• Allocation –Primary –Secondary	• Assignment • Identification
• Allotment	• Designation
• Use	• Gray spectrum

These categories are:

- *Allocation.* An allocation of spectrum means that a discrete portion of spectrum is made available to a specific radiocommunications service. Radiocommunications services include the *fixed service* (FS) (such as wireless cable television), the *mobile service* (MS) (such as mobile telephony services), the *fixed satellite service* (FSS), and the *mobile satellite service* (MSS). Allocations of spectrum, as will be discussed further subsequently in this book, can be made on a domestic, regional, or international basis, depending on the forum in which the allocation occurs. Radiocommunications spectrum can be allocated for a specific service on either a *primary* or *secondary basis.* If spectrum is allocated to a specific service on a primary basis, no other use in the subject frequency band may cause interference to the operation of that service. However, if spectrum is allocated to a specific service on a secondary basis, that service may have to accept harmful interference into its operation by users of the primary service. There may be more than one primary or secondary use authorized in a frequency band. As discussed in subsequent chapters, interference is usually mitigated through compliance with technical operational rules.
- *Allotment.* Allotments are made to areas or countries. In some countries, however, the term allotment when used in a domestic context may have only a very limited meaning for a particular service or station, such as an allotment that is made for an FM radio station. In some countries, allotments may be defined in the same manner as *use.*
- *Use.* Individual countries make determinations on the use of specific frequency bands that have been allocated to a specified radiocommunications service. For example, for a frequency band that is allocated to the mobile service, the individual government will then need to make a determination as to what uses, such as paging or mobile telephony, should be authorized to operate within the frequency band and to what technical rules the spectrum should be subject. Accordingly, uses may be thought of as the specific type of use that may be made of the allocation, as long as the use fits within the broad definition of the allocated radiocommunications service to that band (e.g., 3G services in the MS band). Governments also implement regulations that govern how the relevant frequency band

can be utilized (e.g., through technical and operational limitations and requirements).

- *Assignment* (also known as an authorization, license, or concession). An assignment involves a government grant to a specific applicant and/or use to utilize a portion of the spectrum resource, generally for a specified use with set technical parameters. For example, in the United States, radiocommunications spectrum may be allocated to the mobile service but assigned more specifically to use by the cellular telephony service. The spectrum may then be assigned even more specifically to specified operators, such as BellSouth, so that no other types of mobile services may obtain an authorization to operate in that frequency band without an accompanying rule change. Assignments of the radiocommunications spectrum can be granted to individual applicants by a government in a number of ways. These methods include a straight application processes, comparative hearings, lotteries, and auctions. As discussed throughout this book, auctions have increased in attractiveness for many countries because of the nonsubjective nature of the process and the financial gains for governments that have typically been associated with the larger auctions that have attracted well-financed bidders. However, the ability to continue to attract the financial incentives that have been seen in the past is at jeopardy because of the recently changed telecommunications financial landscape.

- *Identification.* An identification of spectrum is an international concept developed at the ITU that provides guidance to countries internationally on a specific allocation of spectrum that may be used by a large number of countries for a specified use (that is not a category of radiocommunications service). Because the term *identification* is not defined in the ITU Radio Regulations, it has no regulatory status and is not required to be followed by any country; it is simply a guideline. A good example of a recent identification of spectrum was to the use of certain frequency bands allocated by the ITU to the mobile service on a primary basis for 3G services.

- *Designation.* This is another international concept developed at the ITU. A designation of spectrum implies that a specific frequency band should be used by a particular type of system based on the limits imposed by the accompanying technical rules. Designations of spectrum also have no regulatory status under the ITU rules and regulations.

- *Gray spectrum.* A concept that is being expanded more fully by governments and policy makers is the concept of gray spectrum. This approach, which has not yet been adopted either internationally or domestically, would leave large swathes of spectrum available for flexible use by authorized service providers, largely based on what use places the highest monetary value on the relevant frequency band.

These terms and concepts are the cornerstones of spectrum management and are key to understanding spectrum utilization. They shall be relied upon throughout the rest of this book.

A brief background of spectrum management

In order to understand the structure surrounding today's radiocommunications spectrum battles, it is imperative that the background behind today's spectrum management regime be understood. Specifically, it is important to know that that the current regime has evolved from one whereby governments had the primary role in all spectrum use decisions, to one where the private sector has had an increasingly growing and important role. In addition, the increased evolutionary pace of technology has also placed pressure on the traditional operations of the spectrum management regime. This changing dynamic has caused the spectrum management system to adapt to one in which spectrum-use changes occur at a far more rapid pace. In addition, the increasingly congested radiocommunications spectrum resource has also caused alarm in experts who believe that in the future, current spectrum-management forums will be ill equipped to handle the issues that arise.

Although the theoretical basis of radiocommunications spectrum transmission was first formed in 1873 by James Maxwell, it was only in 1888 that Henreich Hertz was able to confirm Maxwell's theory through the generation and transmission of electromagnetic waves [5]. However the practical use of radiocommunications was only demonstrated in 1885 in separate experiments by Alexander Popov and Guglielmo Marconi [6].

In the beginning of the development of telecommunications technology, radiocommunications spectrum was utilized only for limited services, starting with a form of wireless telegraphy [7]. This was primarily aimed at use for ship-to-ship and ship-to-shore communications. Slowly, new wireless services were developed by the public and private sectors, which began to access larger portions and more remote sections of the radiocommunications spectrum resource. Often these services were utilized by pubic-type uses, such as police and fire department communications. Hence, in most countries, the government had almost exclusive control of the radiocommunications spectrum.[1]

As technology improved and the commercial benefits of use of the spectrum became more apparent, private entities increasingly began to recognize the importance of wireless technology as both an adjunct to the existing wireline network, and, over time, as an important standalone transmission technology. This meant that slowly companies became more engaged in obtaining radiocommunications frequencies for private use. In fact, in many cases, radiocommunications-based services were seen as a way to jump start competition in a telecommunications market, because oftentimes they were not the prime means of providing telecommunications services by the incumbent provider. Hence, the threat that was apparent from wireline competitive services was often seen as substantially lessened by entering the market through a wireless service offering. A good example of this has been the *very small aperture terminal* (VSAT) market, whereby one of the first services that is often allowed into a given marketplace is a receive-only data service provided to VSAT terminals for private usage. VSATs are used to provide satellite services to individual users.

As new uses and users of the radiocommunications spectrum blossomed, many spectrum issues began to arise. These issues included what services should be allowed to operate in what frequency band, how to avoid crowded spectrum, how to ensure the continued viability of existing systems, and how to ensure that systems were interoperable. In addition, because of the international nature of the radiocommunications spectrum (e.g., that services could operate across national boundaries and cause harmful interference into other services even on an inadvertent basis), issues of cross-border interference began to arise. This required

1. An exception to the general policy of government control of spectrum was in the United States, which often encouraged private operation of radio services. For example, as technologies permitting efficient, two-way voice communications over radiocommunications spectrum developed, the United States relied on the initiative of service providers to deploy these services to the public.

governments of nearby and bordering states to increasingly coordinate their usage of the radiocommunications spectrum with one another or face harmful interference. Further, equipment manufacturers began to raise concerns about the implementation of some uniformity of use of the spectrum in order to build more standard communications equipment, which could be sold in more than one country (a debate that continues even today).

Governments soon began developing and implementing their own regulations. However, there was a complicating factor: Radiocommunications propagation was not well understood at this time. Radiocommunications propagation, at low frequencies, was greatly affected by unpredictable changes. Radiocommunications waves could provide good service at large distances as long as certain conditions were met. Therefore, a detailed paternalistic form of regulation seemed to flow naturally for the government to protect users from this poorly understood phenomena.

However, many governments in the beginning time frame of regulation began to recognize that they would have to act jointly in managing the radiocommunications spectrum resource or harmful interference between users and authorized services would occur. This particularly came to light in 1902, when the King of Prussia, sailing from the United States, was unable to send a courtesy message to U.S. President Theodore Roosevelt because the ship-to-shore equipment was not able to interconnect. Partially in response to this incident, and because of growing concern about interoperability in general, in 1903 a significant portion of the world's countries participated in a preliminary radiocommunications conference under the auspices of the forbearer to the ITU [8]. This conference was the first international meeting to study the question of whether there should be international regulations governing the radiotelegraph service. This meeting provided the foundation for the 1906 *International Radiotelegraph Conference* (IRC), where the first International Radiotelegraph Convention was signed. The IRC provided the first regulations governing wireless telegraphy [8]. These rules would evolve into what are today known as the ITU Radio Regulations. These regulations (as revised throughout the years) continue to be the cornerstone for the international use of spectrum for communications purposes and will be discussed at length throughout this book.

The first set of IRC regulations was quite narrow and focused solely on the radiotelegraph service. However, in 1927, the IRC regulations were extended to apply to frequency bands for other existing radio services, such as the fixed, maritime and aeronautical mobile, broadcasting, amateur, and experimental [9]. This extension was done to ensure greater efficiency of operation in view of the increase in the number of services using frequencies and the technical peculiarities of each service.

In the 1930s, the radio and wireline side of communications jointly formed the ITU, with overall responsibility for all forms of communication, by wire, radio, optical systems, or other electromagnetic systems. In 1947, the ITU became a specialized agency of the United Nations [10]. The ITU through today has continued the tradition of being the international body for the allocation of radiocommunications spectrum on both a global and regional basis.

One of the initial organs of the ITU was the *International Frequency Registration Board* (IFRB), which was established to manage the radiocommunications spectrum globally [11]. In addition, the IFRB made mandatory the International Table of Frequency Allocations. This table allocates spectrum to each service using radiocommunications spectrum in specific frequency bands with an eye toward avoiding interference between users in the different services. The International Table of Frequency Allocations is still relied upon, utilized, and required to be adhered to by each member state of the ITU.

Over time, the organizational structure of the ITU has changed, largely in response to the quicker pace of technological advances and the influx of new services brought by the increasingly competitive telecommunications marketplace. Today, radiocommunications spectrum issues are managed on a day-to-day basis by the ITU's Radiocommunications Sector. In order to establish radiocommunications spectrum allocations, the members of the ITU meet in *World Radiocommunications Conferences* (WRCs), where changes to the International Table of Frequency Allocations are made and associated technical rules governing use of the allocations are adopted in the form of recommendations, resolutions, and the like [12]. As will be discussed later in this book, the ITU Radiocommunications Sector also provides expert advice in the advance work for these conferences, including holding study group meetings and preparatory conferences.

Until the 1990s in many countries, governments, or government-owned service providers, were the main participants at the WRCs and the accompanying preparatory meetings.[2] Today, as discussed in further

2. A notable exception to this was the United States, where the private sector has been a participant in WRCs and the accompanying preparatory processes for many years.

detail later in this book, private-sector participants have an increasing role at the ITU. They often join the ITU as private-sector members of the association's sectors in addition to serving on the official delegations of member states at treaty-making and other meetings.

More specifically, as elaborated on in subsequent chapters, the ITU Radiocommunications Sector has the mandate to:

- Effect the allocation of frequency bands, the allotment of radiocommunications frequencies, and the registration of both radiocommunications frequency assignments and any associated orbital position in the geostationary satellite orbit in order to avoid harmful interference between radiocommunications stations of different countries;
- Coordinate efforts to both eliminate harmful interference between radio stations of different countries as well as improve the use made of radiocommunications frequencies and of the geostationary-satellite orbit for radiocommunications services [12].

In general, individual member states utilize the International Table of Frequency Allocations as firm guidance on implementing their own domestic allocation schemes. In fact, because the International Table of Frequency Allocations has treaty status, member states are bound by this document. However, under ITU regulation, countries can take specific reservations to the table and institute a nonconforming use. Even a country that takes such a reservation has an obligation to utilize the radiocommunications spectrum in a manner that does not cause harmful interference to uses that are operating in conformance with the table.

Despite these treaty obligations, it is important to note that there are no explicit enforcement mechanisms to force countries to comply with the table or to punish countries that fail to comply. The International Table of Frequency Allocations is primarily successful because of the likelihood of harmful interference ensuing if a significant number of countries do not comply with the table. In addition, by abiding by the Radio Regulations and accompanying Recommendations and Resolutions, countries ensure the efficient use of spectrum, ease of interconnection, and performance quality, and they obtain predictability in operations.

In parallel with the development and advancement of the international process governing spectrum allocation, the domestic processes around the globe also matured. Over time, based largely on ITU Treaty commitments, each country developed domestic spectrum-allocation schemes. These also included assignment of the allocations to more specific classes of services and, ultimately, rules governing the licensing and use of the radiocommunications spectrum.

Because each country has its own domestic Table of Frequency Allocations, users of the radiocommunications spectrum must also expend time and resources to ensure that the countries where they wish to operate will adopt allocation, assignment, and use schemes that ensure that they can operate in their preferred frequency band. This is especially important to satellite service providers, who generally operate on a global or regional basis, and equipment manufacturers, who wish to sell their equipment to global markets.

As discussed, the domestic process is often more protracted and politically charged than the international process, especially when it comes to new allocations and the introduction of new uses in already-utilized frequency bands. However, the international and domestic processes are generally closely tied together, so political success in both forums are closely linked. Typically, domestic spectrum wars occur under the primary jurisdiction of the regulator, but in some countries other political and governmental bodies may be involved. These bodies can include the defense, state, and commerce agencies, as well as members of the executive and legislative branches. The more politically charged an issue, the longer it generally takes to resolve, and the more bureaucratic the process may become.

In addition to domestic allocations, countries need to create assignments for each service in the allocation scheme in a manner that meets the technical limitations agreed to at the ITU. Assignments provide the more specific rules and regulations for use of the spectrum. It is on this basis that authorizations are awarded to individual users of the spectrum resource generally for a set use. This process, as explained subsequently, can be quite contentious and resource intensive.

Companies, governments, and other interests

Many different interests currently operate or plan to operate telecommunications systems in the radiocommunications frequency spectrum. They are willing to spend significant amount of resources, including money, on retaining and obtaining access to critical areas for their operation of the radiocommunications spectrum resource. To appreciate the underpinnings of spectrum battles, it is important to understand these interests.

These interests include governmental, corporate, scientific, and special interests. Essentially there are two basic ways to break these users into groups. The first concerns whether a particular user is currently an *incumbent user*. The second includes the category of user that is using or plans to use the spectrum (*new* or *planned user*). Both of these are discussed next. This distinction, as will be explained in more depth later in this book, is critical to understanding the pressures that are placed on the radiocommunications spectrum resource.

Incumbent and planned/new users

For purposes of this book and understanding the underpinnings of spectrum wars, there are two basic user groups—*incumbent users* and *planned* or *new users*. Incumbent users are the users that are currently operating in the discrete portion of the radiocommunications spectrum or planning to enter a portion of the spectrum that is already being utilized by users in the same or another service or use. Planned or new users are users who are seeking to operate in spectrum that is currently used or planned to be used by another service or use. As allocation, assignment, and ultimately licensing decisions are made, generally both types of users are impacted. This impact will be further addressed as the case studies unfold here and in subsequent chapters. As will be seen, the conflict between incumbent and new or planned users often results in bitter conflict among the categories of users.

The various interests

Users are discussed here by the broader category of the interest of the user, such as commercial interests, governmental interests, and the military interest. As this book develops, it is important to keep each interest in mind, as this often colors the path and the outcome of the spectrum allocation and assignment debates.

The commercial interest

The greatest focus on new and expanded use of the spectrum resource over the past couple of decades has stemmed from commercial interests. The primary reason for this phenomenon is the increasing attractiveness of wireless telecommunications services to consumers. This has been caused by increased demand by consumers as technology has continued to make innovative products commercially viable. Such services have included improving mobile telecommunications services, such as second generation and 3G mobile services, mobile satellite services, and paging services, as well as FSs, such as wireless cable services. For example, the Meta Group expects that by 2004–2005, 65% to 75% of enterprises will deploy mission-critical applications for wireless and/or pervasive platforms and expects that 75% of corporate knowledge workers will be mobile at least 25% of the time [13].

Because of the increasing financial attractiveness of wireless services, service providers, network operators, and equipment vendors place critical importance on ensuring that the allocation and assignment schemes (and ultimately the licensing regime) enable these new services to operate. Often the search for such spectrum results in the displacement of incumbent users from the spectrum or results in alliances between potential competitors. In the former case, this may result in the subsequent reallocation or reassignment of spectrum from one use in order to accommodate the displaced incumbent users. In the latter case, such alliances may dramatically change the dynamics of an allocation battle, even midstream.

However, allocation, assignment, and licensing are not the only ways for the commercial interests to obtain spectrum. As the 2000s progress, it is becoming more evident that many of the providers that expended large amounts of money to access valuable spectrum are not in a position to finance the buildout of their systems or are apt to delay such a buildout. Accordingly, a second market has been developed, where these licensees resell their rights in part or in sum or to seek governmental authorization to utilize the spectrum where they operate or plan to operate for a new use.

A good example of this was the recent proceeding at the *Federal Communications Commission* (FCC) to allow the utilization of spectrum that was originally assigned to the mobile satellite service for mobile terrestrial wireless services [14]. In these situations, the use of traditional spectrum-management tools are relied upon but fortified with new processes that are developed to handle these unexpected issues and new processes. Other new tools are also being explored, such as the development of the concept of gray spectrum, where economics dictate the use of the spectrum.

Government and other interests

Corporations are not the only entities that are willing to spend resources to obtain and retain spectrum usage. Many other groups are also active in this arena. Among the most notable is the military. The military is one of the largest users of the radiocommunications spectrum around the globe and is heavily dependent upon spectrum to ensure its communications needs are met. This dependency is often in direct conflict with commercial interests and has led to some very interesting political battles and solutions.

Other interests include public safety, such as use by the police or coast guard, and the scientific community, which has a special focus on the space services and the radio astronomers.

The military interest

Some of the most vocal and powerful users of the spectrum resource are the militaries of each country and the combined power of allied powers, such as those represented in the North Atlantic Treaty Organization. Because the majority of the spectrum was traditionally controlled by individual governments, military interests had their pick of the spectrum. Often times, the military would retain the right to spectrum but would not use it or would operate in an inefficient manner. Traditional services for which the military has utilized the spectrum include remote sensing, satellite communications, terrestrial wireless communications to ensure communications between troops, and air-to-ground communications. Further, because of the clandestine nature of most military operations, the military establishment is often hesitant to share information about either planned or operating systems with the private sector or with other countries. This has made it difficult for telecommunications regulators to redeploy underutilized military spectrum for other uses.

However, over time, especially in the 1990s, as commercial interests have increasingly extended their reach and interest in the spectrum resource, the military has had to expend its resources, including political clout, in retaining existing spectrum and obtaining new spectrum for more innovative and technologically advanced services. Because of its vast power in most countries, the military has generally had the upper hand in such spectrum wars. This, however, had begun to change in the late twentieth century, most notably in the United States and Japan. In other portions of the world, the military is still directly or indirectly able to control the national spectrum-management program, but this is not the case in the United States or Japan. In Japan, the military is weak because of post–World War II reforms. With regard to the United States, the military finds itself in a weaker position than in most countries because of the separation between the FCC and the NTIA, which regulates the spectrum available to the U.S. government.

This change was in part due to decreasing military budgets. Decreased budgets often mean that the military has to rely on the private sector for at least a part of its communications needs (as opposed to building its own costly networks). Accordingly, the military has less of a need to obtain the authorization necessary to directly access the spectrum. Further, as governments see communications as a bigger and bigger key to raising the gross domestic product of their countries, governments may be more open to supporting private-sector communications initiatives over that of the military, especially where the military use may be seen as either inefficient or nonexistent. It is unknown whether this trend will continue due to the war on terrorism that ignited in light of the events of September 11, 2001.

An example of the scientific community: radio astronomers

The scientific community has traditionally been active in radiocommunications spectrum debates. Perhaps the most notable subset of the scientific community has been the radio astronomers. For example, a large amount of current knowledge of the universe has stemmed from radiocommunications astronomy, including directly or indirectly the discovery of quasars, pulsars, and black holes.

Radiocommunications astronomy is largely dependent upon the use of the radiocommunications spectrum in order to further its study of space. This type of scientific study originally tended to be in less commercially attractive and technically isolated higher frequency bands, but that has begun to change. In addition, the scientific community does have access to some of the more attractive lower frequency bands. For example, in the 1980s the nongeostationary mobile satellite service wanted to utilize the 1.6-GHz band, which is where the radio astronomers operated. The FCC held a rulemaking proceeding in order to facilitate sharing of this frequency band between the two services.

As new consumer and government radiocommunications services and uses are able to operate reliably in higher frequency bands, the radio astronomy service has felt increasing pressure on its continued primary or exclusive use of these frequency bands. It is likely that over the next 10 years, the higher frequency bands where the scientific community has been able to operate in relative peace will be the subject of more frequent and more brutal spectrum wars.

Public safety uses

As discussed earlier, one the earliest uses of the radio frequency spectrum was for safety-at-sea purposes. Today, this is just one of many important uses of the spectrum for public-safety uses. Other such uses can include air-traffic control, police communications, and fire dispatch. Traditionally, governments, intergovernmental organizations, and not-for-profit corporations have been the major advocates for spectrum for these uses.

Today, this spectrum is often in the valuable lower portions of the frequency bands. Accordingly, it is often under attack for commercial use and has had to accept at times relocation to other frequency bands or shared use of a frequency band with other users. The battles over access to spectrum used by this use are often quite contentious because of the public-interest nature of public-safety uses.

Two unique battles for the spectrum resource for new services

What follows is an overview of two interesting case studies where the search and battle for spectrum for new services was intense and often left other parties injured. Each will be referred to in subsequent chapters of this book to illustrate key points. The first of these case studies involves the search for spectrum in an upper portion of the spectrum resource, the Ka or 28-GHz band, for use by the NGSO FSS in the mid-1990s. The second case study is focused on the recent spectrum war over the identification and allocation of spectrum for 3G services, an advanced form of mobile telephony and data communications.

NGSO FSS: taking the world by storm

Until the 1990s, spectrum for satellite use had several types of allocations, including that of the FSS and the MSS. These services envisioned operation by satellites at a *geostationary orbit* (GSO) (approximately 42,164 km above the Earth's equator) [15]. However, in the late 1980s and early 1990s, a new type of satellite service was proposed . This new service, which would provide broadband data and voice services, would utilize a satellite system that operates in a *nongeostationary orbit* (NGSO) (approximately 780 km above the Earth) [16]. It was argued that these NGSO satellite systems would be commercially attractive for broadband services because they eliminated the transmission delay typically associated with communication that traveled the great distance between Earth and the GSO satellite. The minimization of such delay was very important in order to have commercially viable services for voice and high-speed data uses. In order to capitalize on this use, the first proposed NGSO system was that of a MSS NGSO system, primarily advanced by a company called Iridium.

Iridium envisioned an MSS NGSO system that would provide global mobile telephony and low-speed data services (such as paging) to all parts of the globe (using the land-based mobile telephony network as an adjunct) [17]. By the mid-1990s, this service began to look more viable as the ITU at WRC 1992 had allocated spectrum for this new type of service—NGSO MSS—and the FCC was working on several major proceedings on this issue. Unfortunately, shortly after the Iridium system was launched, it went into bankruptcy. Today, the operating entity Iridium has a scaled-back business plan, and many planned competing NGSO MSS systems have not been deployed.

During this time, other private-sector interests were examining whether to use this NGSO technology for other uses as well. One of the most innovative uses that were proposed was an NGSO FSS system that would operate in the microwave bands, one of which is designated the Ka band. Such a system would be able to provide broadband FSS to a global audience without suffering the transmission delay traditionally associated with GSO systems.

However, the Ka band was under scrutiny for use by other services. First, the Ka band was allocated internationally on a coprimary basis to the GSO FSS and the FS. Such an allocation meant that NGSO FSS might not be able to operate in this band because ITU Radio Regulation 2613 provided priority for GSO systems in the band. In addition, the FS was becoming more diligent in using this band, and a new type of FS system was being put into experimental use in the United States and several other countries. This new use, called *local multipoint distribution service* (LMDS), was being pioneered by a company called Cellularvision. The Cellularvision system provided access to both video and broadband services within an approximately 5-km cell [18]. Later, other FS uses of the band would cause further complications to the NGSO FSS because of sharing concerns [19].

In light of these movements, the first major advocate of NGSO FSS, Teledesic Corporation,[3] determined (with some prodding from the FCC)

3. Teledesic Corporation was initially financed by Craig McCaw and Bill Gates.

that it needed to make its intentions known that it wanted to utilize the Ka band for NGSO FSS. Accordingly, in 1994, Teledesic Corporation filed its application at the FCC, opening the door on a new world in spectrum management and allocation issues. Teledesic's application created a wave of press both for its novel proposal (a satellite system with over 800 orbiting satellites to provide broadband services on a global basis) and for its investors, Craig McCaw of McCaw Cellular fame and Bill Gates, the chairman of Microsoft.

Teledesic's application opened a new world of allocation issues at the FCC, at the ITU, and ultimately worldwide. To begin with, the FCC informed Teledesic that it would have to work to make sure that its issues were addressed at the upcoming 1995 WRC. This was a monumental task because the issue of NGSO FSS was not even on the planned agenda for that conference, which was set in 1992. Accordingly, Teledesic worked through the U.S. WRC preparatory process to ensure that the United States would have a proposal to submit for the global designation of spectrum for this use. As discussed in the next several chapters, the U.S. preparatory process involves working within the FCC and with the Department of State and the Department of Commerce on formulating the positions of the United States for the WRC.

Obtaining such a result was not easy. First, many of the other participants of the WRC preparatory process in the United States saw Teledesic as a competitor whom they did not want to succeed. Others, such as the FS users (both current and planned) feared that the Teledesic system would either displace them or that operation of both uses in the same band might cause harmful interference that might result in operational limits being placed on their systems. Of equal importance, many of the government participants were skeptical of Teledesic's motives. It would take Teledesic substantial resources in political clout, the hiring of expert consultants, obtaining key allies through vendor contracts, and the production of technical studies to obtain support for a proposal for this allocation for WRC 1995.

While this U.S. preparatory process was proceeding, Teledesic was also cognizant of the need to obtain international support for the U.S. proposal for the global designation of spectrum for the NGSO FSS. Without international support, the Teledesic proposal would fail at the conference and would have no chance of being deployed. Accordingly, Teledesic worked within the confines of multiple governmental processes to obtain such support. This included setting the ground work through international and regional meetings on the benefits of its proposed system and retaining experts and political consultants in many countries to assist with its advancement. In this regard, the developing world was a key focus, especially because the proposed charges for usage of the Teledesic system was to provide global services and was to be based on a distance-insensitive basis. This would mean that all countries, including lesser developed countries, would have nondiscriminatory access to the Teledesic system. Accordingly, even remote and rural regions would have access to the system.

Teledesic still had its application pending at the FCC. Because of the potential interference between the NGSO FSS and the LMDS service (as well as potential GSO systems in the bands and other FS uses), the FCC initiated a negotiated rulemaking proceeding to see whether the parties could come up with negotiated technical rules so that both systems could operate in the same portions of the Ka band. Over time, it became evident that the parties would be unable to agree upon technical rules for spectrum sharing. Accordingly, the FCC worked with industry to develop a frequency band segmentation plan that provided spectrum for each service to operate within.

In addition, Teledesic had to contend with Norris Communications, an existing GSO system that had received an FCC authorization in the early 1990s to operate in the Ka band. Teledesic felt that because Norris Communications had failed to meet its FCC-mandated buildout obligations for its satellite system, it deserved to have its authorization revoked. Over time, the FCC would agree and ultimately revoked the Norris Communications satellite authorization for failure to build out its system in accordance with its authorization, one of the few times that the FCC had been bold enough to take such action with regard to a satellite system [20].

Further, the FCC had placed the Teledesic license application on public notice for the filing of other systems in that band with which there might be interference and also for any objections. Under the Communications Act of 1934 and the Administrative Procedure Act, the FCC is obligated, except in rare instances, to place applications on public notice to receive any comments or oppositions on the application. This resulted in the filings by multiple proposed satellite providers, using both GSO and NGSO systems. In addition, many users in the FS raised concerns about the potential for harmful interference.

Accordingly, Teledesic had a spectrum war in full progress on many fronts by the time WRC 1995 occurred. Teledesic still did not have an operating authorization. However, because the FCC was considering a band-segmentation plan that provided for use of specific spectrum for the NGSO FSS, Teledesic had won the fight in the United States. The United States was then able to introduce and support the proposal of the creation of a new allocation for NGSO FSS at WRC 95.

A fierce debate ensued at WRC 95 over the proposed allocation of radiocommunications spectrum in the Ka band to the NGSO FSS. Many countries, especially those in Western Europe, were strongly opposed to the allocation of spectrum to the NGSO FSS. This was in part because they did not believe that this service would be successful, and in other part because they feared this service was a new competitor to their existing uses. Further, other countries, such as France, had their own planned NGSO FSS system that operated in another frequency band. Allowing other NGSO FSS systems to operate would prove to increase competition to such a system.

Teledesic, however, was able to gain early support from Israel and some of the developing countries attending the conference. Over time, Teledesic and the United States would expend substantial resources courting the developing world as allies for the NGSO FSS allocation.

As the conference wore on, a compromise solution was reached. WRC 1995 would provide the NGSO FSS with a provisional global designation of 400 MHz of spectrum (100 MHz less than Teledesic had initially sought). This designation would be the subject of further debate at WRC 1997.

In the subsequent conference, in November 1997, WRC 1997 finalized its designation of international radiocommunications spectrum for use by NGSO FSS, such as those Teledesic would provide. In March 1997, the FCC granted Teledesic its license, without regard to the competing applications pending at the FCC. In addition, it soon became apparent that the interference concerns associated with the Teledesic system operation were legitimate. Accordingly, the FCC moved some users to other frequency bands. Teledesic recently announced that it planned to cease construction of its satellite system, blaming the recent collapse of the telecommunications industry [21].

Third generation mobile service and the FS: a compromise

While the wireless market is ramping up on a global scale, many companies have focused their attention on technology that will not be available for another several years in many countries—or until the advent of 3G wireless phones. These 3G devices promise faster and clearer connection rates that will allow users to watch streamed video or listen to streamed audio files through their wireless devices. The end result is that 3G will make the Internet (and voice services) available anywhere via palm-sized mobile phones. The 3G services that have already been released in Japan, and a few other countries, are proving to be fairly popular by consumers.

In order to ensure that sufficient spectrum was available for such services, the 1992 World Administrative Conference (1992 WARC) created the identification of 230 MHz of spectrum for 3G services on a global basis. However, this identification did not result in the use of this band across the globe for this service. To the contrary, the United States and several other countries determined that they had other needs in the identified bands so that the identified bands were not allocated domestically to 3G services.

For example, the United States allocated the 1,900-MHz band to *personal communications services* (PCS). Shortly after this action, the United States came under severe criticism from various international factions, most notably the *European Union* (EU), for what critics argue is a nonconforming use of the IMT 2000 bands. The EU has consistently taken the position that the 1992 initial "identification" was the equivalent to an allocation of spectrum. However, the United States has argued that this argument is legally flawed, on the grounds that the term *identification* has no legal status in the ITU Radio Regulations [22]. The use of the 1992 IMT 2000 bands for nonconforming uses created great concern by many nations, who were once again afraid that there would not be a global identification of spectrum, in the period leading to WRC 2000 [23]. This tension would greatly influence the ability of the United States to reach a consensus domestic position on the identification of additional spectrum for 3G services early in the WRC 2000 preparation process.

It was soon apparent that the initial identification of spectrum would be insufficient to satisfy the spectrum requirements for IMT 2000 systems. Over the next few years preceding WRC 2000, it was determined within the ITU study groups focusing on 3G that a minimum of an additional 160 MHz of spectrum was required for these systems in order to satisfy global IMT 2000 needs through 2010 [24]. In addition, the study groups determined that the most suitable bands for 3G were the 1,710- to 1,855-MHz band (the 1.7-Hz band), the 2,500- to 2,690-MHz band (the 2.5-GHz band), and the 2,700- to 2,900-MHz band (2.8-GHz band) [25]. As discussed next, the current usage of these bands by individual countries and other uses greatly influenced the results of the WRC 2000. One consensus band was going to be hard to find—each of the proposed bands had advocates and adversaries, depending on the current operational use of the band in a specific country or region.

The 1.7-GHz band is allocated to the MS and FS on a coprimary basis under the ITU Radio Regulations [26]. Accordingly, no change to the Table of Allocations was necessary for an identification of spectrum to 3G services. However, the identification of this band had some very fierce and politically powerful opponents, including the Western Europeans and the U.S. *Department of Defense* (DOD) [27]. The DOD would be a strong advocate against utilizing this band for 3G services in both the domestic and international forums.

The 2.5-GHz band was also very controversial. Like the 1.7-GHz band, this band was already allocated under the ITU Radio Regulations to mobile and other services (e.g., the FS, the broadcasting satellite service, and the MSS) [28]. This additional allocation made it very attractive to the MSS community, who foresaw the identification of MSS spectrum for IMT 2000 as a key to their hope of recovering competitiveness internationally. Traditionally, this band had only been used by a burgeoning FS system called *multipoint multichannel distribution systems* (MMDS), a one-way cable alternative. However, in the year or two before the conference, the regulatory landscape had changed and MMDS was poised to be a two-way broadband solution to competitive local access in many countries, including the United States [29].

The final band, the 2.8-GHz band, was not allocated for MS. Accordingly, it was the most controversial at the onset because it would require a change in the ITU Radio Regulations to identify the band for IMT 2000. Compounding this band's unsuitability for IMT 2000 was the use of this band by the U.S. *Federal Aviation Administration* (FAA) and the U.S. Weather Service [30]. Because of these critical applications, this band was not favored by many countries for identification for IMT 2000 and was taken off the table early in the conference.

Because of the divergent interests on all sides of the battle in the years preceding WRC 2000, the result was uncertain as to whether there would even be an identification, much less in which band. The United States was hesitant to take any position on which spectrum should be identified. The EU and its allies strongly supported use of the 2.5-GHz band, while the Americas region (excluding the United States) was strongly in favor of use of the 1.7-GHz band. Many countries believed that the entire conference could be at jeopardy and the validity of any decision would be questioned if the United States did not actively support the decision of the conference.

Accordingly, many governments heavily lobbied the U.S. government to determine which frequency band it supported for identification to 3G.

As WRC 2000 grew closer, it became increasingly apparent in the United States that some action was required in order to respond to the international need for 3G spectrum. The United States convened a group of approximately 15 interested companies and government representatives to negotiate a solution. Ultimately, the proposal that carried the day, as proposed by the MMDS advocates, was a permissive scheme whereby both the 1.7-GHz and the 2.5-GHz bands would be identified for use, with such use subject to each country's own choice. All sides of the debate were equally happy or unhappy with this decision, as it would mean that the domestic processes of allocation and assignment would be heavily relied on.

As WRC 2000 neared, the United States actively advanced this proposal at international and regional conferences, as well as in bilateral negotiations. At first, international support seemed uncertain. However, by the time this proposal was introduced at WRC 2000, it had gained momentum. Over the course of the conference, because of both the extensive lobbying by the United States of other delegations and the chasm that existed between regions supporting either the 1.7-GHz band or the 2.5-GHz band, the proposal was adopted [31].

With the identification agreed to by the conference, the work returned to the domestic governments to determine the appropriate band in which to authorize 3G services. In the United States, the debate appears to have been settled after a protracted proceeding, with the 1.7-GHz band being authorized for 3G and the MMDS community being able to remain within the 2.5-GHz band. In Europe, use of the 2.5-GHz band is preferred.

Demand for 3G services appears to be high. This has been demonstrated by the service providers who have lately entered a pan-European bidding war for 3G spectrum licenses using the 1992 spectrum identification for 3G. The United Kingdom, which led a high-profile auction in the early 2000s, had consortiums bid upward of $6 billion for a *Universal Mobile Telecommunications System* (UMTS) license, which gives holders the right to deploy high-speed networks through 3G spectrum. Other countries, including France and Sweden, have opted to distribute 3G spectrum through a "beauty contest," licensing applicants according to merit rather than available funds.

Critics of the recent furor surrounding 3G devices have questioned the ability of bidders to ever reach the green after sinking huge payments into spectrum allocation. The once-believed ideal, that 3G devices would roll out as a global standard, has also been discarded as competing technologies—*wideband code division multiple access* (WCDMA) in Japan and cdma2000 in the United States, among others—now indicate that the way 3G data is distributed in each spectrum will vary.

This issue will increasingly come to the forefront as additional spectrum, including the frequency bands identified for use for 3G at WRC 2000, are allocated and assigned to users throughout the world. This will be further explored throughout this book.

Conclusion

This chapter provided a basis for exploring the many issues surrounding access to the spectrum resource, including a brief overview of the key terms, institutions, and processes involved. Chapter 2 will supplement this discussion with a more in-depth discussion of spectrum in general and will be followed by an extensive discussion of the domestic and international processes governing the radiocommunications spectrum, including a discussion of the key issues involving access to this resource.

Endnotes

[1] Transcript of Remarks of Chairman Michael K. Powell, before Cellular Telecommunications Internet Association's CTIA Wireless 2001, March 20, 2001, Las Vegas, NV, available at http://www.fcc.gov/Speeches/Powell/2001/spmkp101.html.

[2] http://www.tua.co.uk/snipstext.htm.

[3] http://www.spectrumauctions.gov.uk/auction.

[4] See Statement by the International Telecommunication Union at the Third United Nations Conference on the Exploration and Peaceful Uses of Outerspace, available at http://www.un.org/events/unispace3/speeches/20itu.htm.

[5] James Maxwell at http://www.phy.hr/~dpaar/fizicari/xmaxwell.html.

[6] See Guglielmo Marconi at http://www.webstationone.com/fecha/popov.htm.

[7] See Allison, Audrey, "Meeting the Challenges of Change: The Reform of the International Telecommunication Union," *Federal Communications Law Journal*, Vol. 45, 1992, pp. 491–497.

[8] See Allison, Audrey, "Meeting the Challenges of Change: The Reform of the International Telecommunication Union," *Federal Communications Law Journal*, Vol. 45, 1992, pp. 491–498.

[9] See Allison, Audrey, "Meeting the Challenges of Change: The Reform of the International Telecommunication Union," *Federal Communications Law Journal*, Vol. 45, 1992, pp. 491–499.

[10] See Allison, Audrey, "Meeting the Challenges of Change: The Reform of the International Telecommunication Union," *Federal Communications Law Journal*, Vol. 45, 1992, pp. 491–501.

[11] See Allison, Audrey, "Meeting the Challenges of Change: The Reform of the International Telecommunication Union," *Federal Communications Law Journal*, Vol. 45, 1992, pp. 491–502.

[12] See Allison, Audrey,"Meeting the Challenges of Change: The Reform of the International Telecommunication Union," *Federal Communications Law Journal*, Vol. 45, 1992, pp. 491–504.

[13] See "Building the Net: Trends for a Digital Future," at http://www.trends report.net/wireless/6.html.

[14] http://www.fcc.gov/Bureaus/International/News_Releases/2001/nrin0113.htm.

[15] See geostationary orbit definition at http://www.its.bldrdoc.gov/fs-1037/dir-017/_2456.htm.

[16] http://www.idrc.ca/acacia/studies/ir-jens5.htm#Non-Geostationary%20Orbit%20(NGSO)%20Satellites.

[17] Whalen, David, "Communications Satellites: Making the Global Village," at http://www.hq.nasa.gov/office/pao/History/satcomhistory.html.

[18] Bunn, Austin, "Cellularvision Offers Wireless Net in New York," 1997, available at http://www.wired.com/news/topstories/0,1287,3510,00.html.

[19] http://www.ee.surrey.ac.uk/Contrib/Edupage/1997/03/13-03-1997.html#3.

[20] Norris Satellite Communications, Inc., Memorandum, Opinion and Order, 12 FCC Rcd 22299, 1997.

[21] Sharon, Pian Chan, "The Birth and Death of an Idea: Teledesic's 'Internet in the Sky'," *The Seattle Times*, October 7, 2002.

[22] See "Backgrounder: Spectrum for Third Generation IMT-2000 Systems," *World Radiocommunication Conference 2000* (explaining WARC 1992 identification).

[23] See Peichel, Cory, "From Watson to W-CDMA: How Wireless Technologies Evolved: Special Focus: Technology Information," *Communications News*, Vol. 325, No. 5, p. 62 (noting divergent identifications and that the FCC in 1993 allocated the 1.9-GHz band to be auctioned off for PCS).

[24] See Silva, Jeffrey, "3G WRC Policy Dispute Erupts," *Radio Communications Report,* July 5, 1999, p. 1 (citing U.S. draft proposal that states that WARC 92 identifications "do not constitute an allocation and lack definition and regulatory purpose").

[25] See Sidall, David R., "Debate Swirls Around IMT-2000," *Radiocommunication Report,* September 21, 2000, p. 20.

[26] Article S5 of the Radio Regulations, Footnotes S5.339, S.403, S.409–411, S5.413, S.415, S.415A, and S.416–418.

[27] See Huber, Josef F., Vice Chairman, UMTS Forum, "IMT-2000 Spectrum—Views from the UMTS Forum," *World Radiocommunication Conference 2000.*

[28] Article S5 of the Radio Regulations, Footnotes S5.149, S5.341, S5.380, and S.5.385–88.

[29] Whitely, Christopher, "Fixed Wireless Won't Move Unless Carriers Tout Pluses," *Electronic Engineering Times,* November 8, 1999, p. 83.

[30] See "U.S. Sees Spectrum Proposal as 'Bridge' at Upcoming Conference," *Communications Daily,* March 20, 2000; see also Article S5 of the Radio Regulations, Footnotes S5.337 and S.5.424 (allocating 2.7- to 2.9-GHz band for aeronautical radionavigation and weather reporting).

[31] Schoettler, Ambassador Gail, "Fighting for Our Air Waves," *The Denver Post,* March 5, 2000, p. G-3; see also "U.S. Offers Draft Plan for Next-Generation Spectrum Services," *Communications Daily,* February 18, 2000; see also Final Acts of WRC 2000.

2

Spectrum Primer

The underlying rationale for almost every spectrum war is technical. The primary reason that spectrum wars occur is the very real technical limitations that surround the use of the radiocommunications spectrum. Of course, each battle also has very important political, legal/regulatory, and economic rationales, which will be explained throughout this book. However, the underlying controversy in almost every instance is caused by technical concerns. Accordingly, in order to understand how to resolve battles over the use of the radiocommunications spectrum, and the spectrum allocation and assignment process, it is important to have a basic understanding of the technical considerations that are associated with the use of the radiocommunications spectrum resource.

This chapter endeavors to set forth a brief overview of the technical characteristics of the radiocommunications spectrum resource, the issues surrounding spectrum scarcity and harmful interference, and a brief overview of some of the most heavily utilized and commercialized radiocommunications services to which spectrum is currently allocated. This chapter, however, is simply an overview of an extremely complex subject matter. There are many issues that are not discussed, such as link budgets, signal strength issues, and noise and data capacity limits [1].

This chapter concludes with a brief discussion of some of the key considerations that an advocate of a new service or use must examine when determining what spectrum it would like to seek authority for that service or use to operate within. Overall, this chapter establishes the basic groundwork to understand the technical underpinnings of almost all spectrum battles.

Overview of technical characteristics of the radiocommunications spectrum resource

The radiocommunications spectrum is a resource that is made up of radiocommunications waves that operate below 3,000 GHz, though most communications uses utilize spectrum below 400 GHz [2]. The allocated radiocommunications spectrum is located between 9 kHz and 300 GHz. A good example of the realm of operations in the spectrum resource is contained in the spectrum chart in Table 2.1.

Each radio wave is an oscillating electromagnetic wave characterized by frequency and strength. The frequency is a measure of the number of times per second a wave oscillates or cycles per second (hertz). The strength is a measure of the amplitude of the wave or the power (watts). These waves are radiated through free space by a transmitting antenna where the frequency of the wave is proportional to the size of the antenna. Similarly, these waves moving through space can be “caught” by an antenna or receiver that is designed for that frequency and is within the area designated for that signal. Antennas such as those in AM/FM radios or televisions are designed so that several frequency ranges (channel or bandwidth) can be picked up or tuned into. Some antennas, such as mobile telephones, are designed to both transmit and receive radiocommunications waves.

The minimum distance between the transmitter and the receiver in a vacuum is determined by factors such as the specific frequency band and the power of both the transmitter and the receiver. In reality, natural and manmade obstacles, such as buildings, trees, and design imperfections, as well as absorption constraints associated with certain frequencies, must be taken into account. Some of the physical or absorption constraints include:

- Frequencies below 50 MHz, which are directly affected by the ionosphere;

TABLE 2.1
U.S. Spectrum Chart

BANDWIDTH DESCRIPTION	FREQUENCY RANGE
Extremely low frequency (ELF)	0 to 3 kHz
Very low frequency (VLF)	3 kHz to 30 kHz
Radio navigation and maritime/aeronautical mobile	9 kHz to 540 kHz
Low frequency (LF)	30 kHz to 300 kHz
Medium frequency (MF)	300 kHz to 3,000 kHz
AM radio broadcast	540 kHz to 1,630 kHz
Travelers information service	1,610 kHz
High frequency (HF)	3 MHz to 30 MHz
Shortwave broadcast radio	5.95 MHz to 26.1 MHz
Very high frequency (VHF)	30 MHz to 300 MHz
Low band: television band 1 (channels 2–6)	54 MHz to 88 MHz
Mid-band: FM radio broadcast	88 MHz to 174 MHz
High band: television band 2 (Channels 7–13)	174 MHz to 216 MHz
Super band (mobile/fixed radio and television)	216 MHz to 600 MHz
Ultra-high frequency (UHF)	300 MHz to 3,000 MHz
Channels 14–70	470 MHz to 806 MHz
L-band	500 MHz to 1,500 MHz
PCS	1,850 MHz to 1,990 MHz
Unlicensed PCS devices	1,910 MHz to 1,930 MHz
Superhigh frequencies (SHF) (microwave)	3 GHz to 30.0 GHz
C-band	3,600 MHz to 7,025 MHz
X-band	7.25 GHz to 8.4 GHz
Ku-band	10.7 GHz to 14.5 GHz
Ka-band	17.3 GHz to 31.0 GHz
Extremely high frequencies (EHF)(millimeter wave signals)	30.0 GHz to 300 GHz
Additional fixed satellite	38.6 GHz to 275 GHz

Table 2.1 (continued)

Bandwidth Description	Frequency Range
Infrared radiation	300 GHz to 430 THz
Visible light	430 THz to 750 THz
Ultraviolet radiation	1.62 PHz to 30 PHz
X-rays	30 PHz to 30 EHZ
Gamma rays	30 EHZ to 3,000 EHZ

Note: Radio frequency bandwidth: The allocated radiocommunications is located between 9 kHz and 300 GHz.

- Frequencies around 1 GHz, which place constraints on the ability of the communications to go around corners or penetrate buildings;
- Frequencies above 5 GHz, which are affected by rain;
- Frequencies around 60 GHz, whereby communications get absorbed by oxygen in the atmosphere and face other negative impacts.

These properties play critical roles in designing communications systems (for example, playing a role in how close cellular towers are sited or where in your home you install your direct-to-home satellite television dish) when deciding where in the spectrum these systems will operate. For example, in the United States, the following technology operates in the identified frequency bands:

- Garage door openers and alarm systems operate at around 40 MHz.
- Standard cordless phones operate at 40 to 50 MHz.
- Wildlife tracking collars operate at 215 to 220 MHz.
- Cell phones operate at 824 to 849 MHz.
- Air traffic control radar operates at 960 to 1,215 MHz.
- Global positioning systems operate at 1.2 to 1.75 GHz.
- Deep space radiocommunications operate at 2,290 to 2,300 MHz [3].

The width of the radiocommunications waves (bandwidth) required for communications is different for varying types of communications. For example, voice telephony requires 4 kHz of minimum required bandwidth, while NTSC (analog) television requires 6 MHz. The bandwidth of the signal to be sent has, to a great extent, the effect of constraining the frequency that is utilized for the service. This is because traditional radiocommunications systems tend to work optimally if the bandwidth of the signal is less than a few percent of the center frequency band.

Accordingly, use of frequency bands is not random and is developed with an understanding of the technical characteristics of each frequency band. The use of each frequency band by a specific radiocommunications service is determined through both domestic and international processes, which result in the allocation and assignment of spectrum for different services and uses, and the adoption of accompanying technical rules. These processes are often quite contentious and costly. Further, each use has a spectrum requirement in terms of the amount of spectrum it must use to operate. For example, the advocates of 3G spectrum set forth a requirement of a minimum of 160 MHz of contiguous spectrum in which to operate [4]. The amount of spectrum is often as contentious as the placement of the service in a specific frequency band in the radiocommunications allocation table.

Spectrum scarcity and harmful interference

Two basic technical considerations that always impact the use of the radiocommunications spectrum by new services and uses are scarcity and harmful interference. These two concepts must be considered hand in hand. Although spectrum is a limited resource to begin with, its scarcity is further increased because of the need for communications to be free from harmful interference from other radiocommunications services. Some critics believe that scarcity of the radiocommunications spectrum is based on the fact that existing users have little or no incentive to improve the efficiency of their use and that governments do not have the political will to require them to do this. Accordingly, users of the spectrum resource may not operate as efficiently as they can, leading to spectrum congestion and, ultimately, scarcity.

One reason that radiocommunications spectrum is scarce is because it is a limited resource. Today, communications devices are generally only capable of operating in spectrum up to 400 GHz (the ITU Radio Regulations apply only to frequencies between 9 kHz and 400 GHz), with the bulk of communications uses occurring below frequencies in the 30-GHz range.

Accordingly, this scarcity will be lessened as technology is further developed to access a broader range of spectrum [5].

Another important concept that impacts the availability of the radio communications spectrum resource is the potential for harmful interference by one radiocommunications service into another radiocommunications service. A key goal of radiocommunications frequency management is the avoidance of harmful interference. Harmful interference is defined by the ITU Radio Regulations as [6]:

> Interference which endangers the functioning of a radionavigation service or of other safety services or seriously degrades, obstructs or repeatedly interrupts a radiocommunications service operating in accordance with the ITU Radio Regulations.

Under the ITU Radio Regulations, "harmful" interference by one radiocommunications service into another radiocommunications service is prohibited. However, a certain amount of interference into one radiocommunications service by another radiocommunications service is allowed. Interference is defined as [6]:

> The effect of unwanted energy due to one or a combination of emissions, radiations, or inductions upon reception in a radiocommunication system, manifested by any performance degradation, misinterpretation, or loss of information which could be extracted in the absence of such unwanted energy.

The amount of interference that is allowed in a particular frequency band is defined within the ITU Radio Regulations and the Recommendations. In actuality, however, most domestic regulators adopt much stricter guidelines. For example, when acting on the Northpoint Technology application for MVDDS, the FCC placed certain EPFD limits on its use at existing direct broadcast satellite sites [7]. In essence, however, harmful interference between users occurs when the interference is such that it causes serious detrimental effects, such as outages. This is inapposite to interference that is merely a nuisance or that can be overcome by appropriate measures.

In order to avoid harmful interference between radiocommunications services, the ITU and domestic regulators have adopted strict frequency allocation schemes [7]. These spectrum allocations put in place technical guidelines or rules governing the use of specific frequency bands by individual or multiple services so that harmful interference does not occur within a single frequency band or between authorized services and uses operating in adjacent frequency bands.

Services or uses operating in the same band are commonly known as cofrequency sharing. Cofrequency sharing is the common use of the same frequency by two or more services where the potential for interference exists. These services may operate on a coprimary, or primary and secondary basis, or cosecondary basis, depending on the allocation provided for in the relevant Table of Frequency Allocations. In all such cases the primary service has the right for operation free from harmful interference from the secondary service, and the secondary service cannot claim protection from the primary service (although coprimary and cosecondary services are given different priorities generally based on the day they are brought into use). For any frequency-sharing scheme to be successful, complex technical rules often need to be instituted and adhered to by the users of the spectrum. For example, in order to accommodate the fixed wireless access service and the digital audio radio satellite service in the United States and Mexico, the two countries agreed to emission limits among the services on a bilateral basis, above that required by the ITU Radio Regulations [8].

As the Mexican-U.S. example demonstrates, the acceptance of levels of interference between radiocommunications services can be agreed to by countries. Accepted interference is interference that is agreed to by two administrations but is higher than that permitted by the ITU Radio Regulations. This type of arrangement is commonly entered into by countries that have common borders but may have different allocations schemes that allow uses that may interfere with one another.

In order to further guard against interference, WRCs often adopt assignment or allotment plans. Assignment plans involve the assignment of frequencies to each station of a country. Allotment plans are utilized in an effort by the developing and the developed world to ensure that there is available spectrum for their use for critical satellite communications services. When plans are drawn up, suitable technical data, planning parameters and criteria for sharing with other services, as appropriate, are also adopted. All cases of unacceptable interference are resolved in advance and suitable procedures laid down for the bringing into use of the planned frequencies. In each case the bringing into use must be in conformity of the Table of Frequency Allocations and other provisions of the Radio Regulations. When such plans are not in place, international coordination procedures are established under which a country that plans to use a frequency must obtain the agreement of all the countries that might be affected.

Further, countries may be able to obtain additional protection by registering their use of a frequency band in the Master International Register of Frequency bands. Depending upon the service that they are operating, countries may or may not utilize the Master International Frequency Register. Global systems, such as international satellite systems, generally register their use. In regard to terrestrial uses, this determination is made on whether the new use or station is likely to cause interference outside of the territory of the country in which is located. If this is the case, the country where the station is located is required to send a notice to the ITU Radiocommunications Bureau containing the relevant technical standards of the station.

The Radiocommunications Bureau examines the notice to ensure that the proposed use is operating in conformance with the ITU Table of Frequency Allocations and the Radio Regulations [9]. To the extent the Radiocommunications Bureau determines that the station may cause harmful interference to any other stations already notified to the Master Frequency Resister, the proposed use must undergo changes so as to prevent harmful interference. Such changes can include basic technical changes, such as different emission standards, or a change of use of frequencies. If interference is not anticipated, the Radiocommunications Bureau will issue a favorable finding and enter the station in the Master Frequency Register.

The allocation scheme

In this section we focus upon the importance of the allocation scheme from a technical perspective. This will be complimented by the next few subsequent chapters, which discuss the regulatory issues associated with the allocation scheme. The focus of this section is to provide an overview of some of the key radiocommunications services to which spectrum is allocated.

In order to most efficiently manage the radiocommunications spectrum, the ITU has allocated different segments of the radiocommunications spectrum to over 40 different services through the WRC process [10]. These services generally refer to broad types of radiocommunications services that may operate in the spectrum. For example, a common use of the spectrum is an allocation for the mobile service. The mobile service may encompass a wide variety of different uses and technologies, such as pagers or mobile telephones. Many times the accompanying technical rules contained in footnotes to the International Table of Frequency Allocations, or in Resolutions or Recommendations may be constraining in terms of the type of use that may be made of the spectrum and the technical limits imposed on that service. For example, certain emission standards may be imposed on the technologies operated in a frequency band allocated for MS uses. These rules are often enacted to prevent harmful interference among users but also have the direct impact of shaping the specific use that may operate in the frequency band, such that data services but not voice services, can be provided by the service provider.

Some of the more common radiocommunications services to which spectrum is often internationally and domestically allocated—and heavily commercialized or utilized—include:

- *FS:* A radiocommunications service between fixed points. MMDS communications are a form of fixed service.
- *FSS:* A radiocommunications service between Earth stations and satellites. The Earth stations are within a fixed area. The Teledesic satellite system is an NGSO FSS system.
- *MS:* A radiocommunications service between mobile stations or mobile stations and land stations. A form of mobile service is 3G.
- *MSS:* A radiocommunications service between mobile Earth stations and one or more satellites or between satellites used for this service. For example, the Iridium satellite system is a system that operates in the MSS.
- *Aeronautical fixed service (AFS):* This is a radiocommunications service between specified fixed points provided primarily for the safety of air navigation and for the regular, efficient, and economical operation of air transport.
- *Aeronautical mobile service (AMS):* This is a mobile service between aeronautical stations and aircraft stations or between aircraft stations. This may include emergency communications. This service is broken into both safety and nonsafety uses.
- *Aeronautical mobile satellite service (AMSS):* This service is similar to the aeronautical mobile service, but utilizes satellite communications. This can be used for safety or nonsafety services.

- *Earth exploration service (EES):* A radiocommunications service between Earth stations and space stations (or solely between Earth stations) that involves information relating to the characteristics of the Earth, such as the environment. Remote sensing uses are an example of this service.
- *Space research service:* A radiocommunications service in which spacecraft are utilized for scientific and research purposes. A good example of the use of this service is the communications carried between the U.S. *National Aeronautic Space Administration* (NASA) and the International Space Station.

Individual countries will have allocation schemes that also designate spectrum for specific uses. For example, many regulators will establish a specific allocation in the FS and an accompanying designation for the local multipoint distribution service or another use of the spectrum.

Key technical considerations when evaluating spectrum use

Determining the most suitable frequency band for a particular radiocommunications service from a technical perspective is complicated. Many key considerations must be examined to assist in the determination of the optimum frequency band in which a service or use should be operated. This technical review is among the most important of the considerations for moving forward on a plan of action for seeking to obtain an allocation or assignment of spectrum. It may, however, have to be adjusted to take into account political, regulatory, and economic considerations. Accordingly, some advocates for spectrum for new uses or services often have contingency plans outlining other spectrum where they could operate, how much spectrum is needed, and what technical issues they can agree to be flexible upon.

There are many key issues that should be evaluated in any technical review of what spectrum in which a new use or service should seek to operate. Among the most important of these considerations are:

- *The proposed frequency band for operation.* This is the most important technical consideration and may include identifying multiple frequency bands if this is a use that has several components to its operation. It is imperative to locate the optimal band for operation of the service or use. However, because of the relative congestion of the overall radiocommunications spectrum resource, the spectrum advocate may also determine other, less optimal frequency bands (depending on where the advocate is in the design phase) that are available for use. This is in case the advocate needs to compromise on this issue and must develop a plan for operations in other than the preferred frequency band.

- *The cost and delay in obtaining access to a specific frequency band.* Very few spectrum advocates are able to spend significant time or money in gaining access to the radiocommunications spectrum. Accordingly, it is imperative that the spectrum advocate determines how much cost and delay it is willing to commit to and how to minimize the cost and delay. Its access plan must take into account any such constraints and have a solution for addressing these issues.

- *The amount of spectrum required for the use.* In order to determine the amount of spectrum required for the use, one must first determine both the minimum and the maximum amount of spectrum that is required for successful operation of the use. This evaluation should include an analysis for all aspects of a system as well as the potential for growth of the use of the service. For example, for a satellite system, this will include the spectrum for the downlink, uplink, any intersatellite links, and spectrum for components such as telemetry and control. Each of these components may need to operate in separate or the same frequency band, and this should also be delineated in the analysis.

- *The current allocation and uses of the frequency band.* It is imperative that an advocate for a spectrum use analyzes the current allocation and current utilization of the frequency band in which it plans to operate. In some cases, it may be very difficult to change an existing allocation or add a new use in the frequency band. Such was the case when the 3G advocates had proposed identifying the 2.8-GHz band for use by 3G services. The 2.8-GHz band was used for aeronautical safety services. Many countries, including the United States, did not want to take the political or safety risk of allowing in a use that may cause interference into such a critical use of the spectrum or might necessitate its relocation. Accordingly, at WRC 2000 this frequency band was taken off the table for consideration in the very early days of the conference. Astutely recognizing this potential issue, the 3G advocates had also identified two other frequency bands for possible identification, and over the course of the conference they moved the focus of the proposed identification to these two other frequency bands.

- *The propagation issues associated with use of a particular frequency band.* How well a radio wave propagates in a particular frequency band for a particular use is variable. Conditions such as rain, the presence of leaves, and the time of day all may affect the propagation characteristics of the radiocommunications service. If a particular service is particularly impacted by propagation characteristics such as rain, the service level may degrade below an acceptable level for the specified use. Competitors and adversaries are likely to raise negative propagation issues as a reason not allocate spectrum to the specific service. Propagation concerns were a very big issue in the domestic U.S. proceeding over the use of the 28-GHz band for LMDS versus NGSO FSS. Some opponents of the LMDS use argued in the FCC's negotiated rulemaking proceeding and in subsequent proceedings that the rain fade associated with use of the 28-GHz band for LMDS would result in unacceptable performance levels. Accordingly, these LMDS opponents argued that because the service would not perform at an acceptable level, allocation of any part of the 28-GHz band for this use would be a mistake. Over time, the LMDS operators were able to convince the FCC that this argument was flawed and the negative impact on the service from rain fade was acceptable.

- *The signal strength required for reliable service.* Technical limits are often placed on the use or allocation of a frequency band to a specified service. Accordingly, it is imperative that the advocate understands the strength of the signal that is required for the communications service and whether such a signal strength may cause interference into other services or uses already in the band or in adjacent bands. If so, the advocate should look for solutions to protect both itself and the other uses from interference or find other solutions, such as relocation of the other service or technical limitations on emissions.

- *Relative amount of radiocommunications and other electrical interference likely to be encountered.* It is much easier to argue that a new use or service should be authorized in a frequency band if it is not likely to cause or face harmful interference by other services and use in the frequency band. If harmful interference is to be encountered, it is important for the advocate to fully understand the extent of this interference and be able to address it through proposed technical rules or other mechanisms, such as proposals to move the offending use or service from the frequency band.

- *Upper practical limits of the useful radiocommunications frequency spectrum and, in general, what higher limit can be expected in the future due to technological advances.* Any evaluation of use of a frequency band should be forward looking. Therefore, it is important to perform a technical due diligence on both current and possible future uses of the frequency band and address any issues that may arise from such uses in order to ensure its continued availability.

- *The cost differentials of utilizing different frequency bands.* Depending on the frequency band that is planned to be utilized, different costs may be associated with the use. The first is the cost of the network to access the spectrum. Use of higher frequency bands, for example, may require the use of more sophisticated and expensive technology than the utilization of a lower frequency band. This may also involve more expensive user equipment. Further, use of different frequency bands may have other associated costs. For example, if the preferred frequency band has incumbent users, a proposed user may have to agree to pay for relocation for the incumbent uses. Such costs may directly impact the ability of the advocate to meet its proposed business plan.

- *Operating characteristics of transmitters and receivers, including practical limitations, that is, size, cost, and technical characteristics.* Early in the development of the proposal for use of a frequency band, advocates should ensure that they have a good understanding of the communications equipment they plan to utilize for their service. The operational characteristics of this equipment will be heavily scrutinized by regulators, competitors, and other users of the frequency band for efficiency and other characteristics.

Based on an evaluation of these characteristics, the advocate may determine that a single frequency band or multiple bands can be used for the service. A proposed use is generally in a better position if it appears that multiple frequency bands are available for the use. This will allow the advocate to examine which band to seek based on both technical and practical considerations.

Once this review is completed, the advocate is in a good position to develop the technical, regulatory, and political strategy for its efforts to gain access to the frequency band and the amount of spectrum required for its proposed use. The development of this strategy will be discussed in forthcoming chapters.

Other considerations

No matter what technical analysis is produced, it is all but certain to fall under strict and critical technical review by governments, other advocates and competitors, and other users of the radiocommunications spectrum resource. Depending on the forum or forums in which the search for spectrum is occurring, this may include ITU review at WRCs and technical meetings and in domestic regulatory proceedings. Technical reviews are often conducted over long periods of time and can be the subject of much debate among interested parties. Accordingly, the result of these group studies often is a consensus recommendation or opinion and may not necessarily be the optimum technical solution.

For example, the amount of spectrum required for 3G systems was studied quite extensively in the ITU in Study Group 8F [11]. One of the major issues that was examined by the parties was the amount of spectrum that was required by these systems to successfully operate. Many different interests participated in this process, with many having different views on the necessary amount of spectrum. Ultimately, a consensus position of the study group was 160 MHz of contiguous spectrum and was utilized as the technical basis on which to identify spectrum for 3G services at WRC 2000. To this day, many of the advocates of 3G argue that this amount of spectrum was far from optimal for their systems to operate. However, other interested parties have made strong technical showings demonstrating that the 160 MHz of spectrum was too large a number. This technical requirement of 160 MHz of spectrum heavily influenced the frequency bands in which the 3G advocates lobbied in favor of identification, as few frequency bands that had the requisite MS allocation were sufficiently large.

In all cases, the advocate must be prepared to defend its position in this process and be aware of where it can and cannot compromise. In some cases, the objections may be purely technical. However, in at least an equal number of cases, it is likely that objections will be based on either political or economic reasons. In each case, it is imperative that the advocate be prepared to respond to such objections in a timely manner. At times, this may take the form of working outside the technical arena and either raising the issue to a political level or seeking support from similarly situated parties or parties that share a similar interest. This approach was well utilized by the 3G service providers throughout their search for additional spectrum in 2000. These providers were well supported at both domestic and international meetings by the equipment manufacturers who hoped to obtain the business of the service providers for this use. In addition, the service providers also sought high legal government official support for their proposals. They received such support largely because of the strong economics that they demonstrated were associated with the provision of 3G services. These approaches will be discussed throughout the subsequent chapters of this book.

Endnotes

[1] For a broader discussion on this subject, see, for example, http://www.itu.org.

[2] See "Mobile Radio Glossary" at http://www.cs.berkeley.edu/~randy/Courses/CS294.S96/glossary.html; see also http://www.itu.int.

[3] 47 C.F.R. Part 2 (2001).

[4] See various papers at http://www.imt-2000.org/portal/search_items.asp?SearchString=&Topic=IMT-2000&Category=About%20IMT-200&SubCategory=What%20is%20IMT-2000&Type=&Action=Search.

[5] See Reed, David P., "How Wireless Networks Scale: The Illusion of Spectrum Scarcity," presented at the FCC Technology Workshop, April 2002, available at http://www.seas.gwu.edu/~cjackson/TAC/Spectrum%20capacity%20myth%20FCC%20TAC.pdf; see also Leopold, George, "Kennard Again Warns Against Spectrum Scarcity," *EE Times*, October 6, 2000, available at http://www.eetimes.com/story/OEG20001006S0042.

[6] See ITU Radio Regulations, available at http://www.itu.int; see also http://myweb.tiscali.co.uk/rdnronald/rr_s1.htm.

[7] http://www.fcc.gov/Speeches/Powell/Statements/2002/stmkp208.html.

[8] See FCC, "Statement of William Kennard Regarding Mexico DARS Agreement Signing," July 26, 2000, available at http://www.fcc.gov/Speeches/Kennard/Statements/2000/stwek062.html; see also U.S. Department of State, "U.S-Mexico Sign Agreement for Digital Audio Radio Satellite Service," July 26, 2000, available at http://usinfo.state.gov/topical/global/ecom/00072601.htm.

[9] http://www.itu.int/ITU-R/information/brochure/br/index.html.

[10] See ITU Radio Regulations, Vol. 1.

[11] http://www.itu.int/ITU-R/study-groups/sg/sg8/wp8f/index.html.

3

Radio Communications Spectrum and Telecommunications Players

The purpose of this chapter is to explore the various factors that are considered in utilizing a wireless solution with which to provide telecommunications services and some of the key issues associated with the use of the radiocommunications spectrum. In order to achieve this, it provides an overview of the advantages and disadvantages of using a wireline versus a wireless network for telecommunications services. This chapter then delves into the major participants involved in the spectrum forums, including telecommunications operators and users and equipment manufacturers. Ultimately, it provides an additional basis for understanding many of the issues raised in subsequent chapters, including access and use of the radiocommunications spectrum resource.

Wireless versus wireline network solutions

Wireline and wireless telecommunications networks both have certain advantages and disadvantages, both in general and when examined as technical solutions for the provision of specific telecommunications services.

In this section, and as outlined in Tables 3.1 and 3.2, we examine the general advantages and disadvantages of both types of telecommunications networks.[1]

The advantages associated with wireless networks include:

- *Mobility.* Unlike with the fixed network, wireless technologies provide the user with the ability to be mobile while using wireless telecommunications devices [1].
- *Geographic reach.* The ability to reach large numbers of people and cover large geographic distances (including into outer space) with limited infrastructure.
- *Lower costs due to less network equipment.* In many cases, communications services that utilize the radiocommunications spectrum are lower in cost than landline services because of the less resource-intensive network deployment [1].
- *In many cases, the ability to avoid large up-front payments for network building.* With regard to nonsatellite-based networks, wireless service providers are able to build out their networks with less investment. This is because nonsatellite-based wireless networks can start with a smaller coverage area that can be easily and quickly expanded as the network grows. This is in contrast to the wireline network, which requires close to full-scale buildout on day one of operations. Satellite systems, however, are more akin to wireline services, because of the large up-front investment required in the satellite itself [1].
- *Quick deployment.* Wireless networks can generally be deployed on a fast basis because of the limited network requirements (i.e., no extensive wiring). For example, in emergency situations, wireless networks are easily brought to the required service area and deployed. A good example of this were the emergency networks that were deployed on September 11, 2001, to help during the

1. However, it is important to bear in mind that a separate analysis on the benefits or disadvantages of any specific solution would need to be service and technology specific.

TABLE 3.1
Advantages and Disadvantages of Wireless Networks

ADVANTAGES	DISADVANTAGES
Mobility	Propagation concerns
Geographic reach	Interference potential
Lower costs	Expense of regulatory fees
Avoidance of large up-front payments	At times, reliance on wireline network
Quick deployment	
Anytime, anywhere communications	
Less anticompetitive concerns	
Less regulation	
Ability to supplement wireline network	

TABLE 3.2
Advantages and Disadvantages of Wireline Networks

ADVANTAGES	DISADVANTAGES
Reliable communications	Need for imbedded infrastructure
Elimination of interference concerns	Fixed service only
Decrease in network cost as use increases	Slow deployment

emergencies at the World Trade Center, where the landline network was severely damaged [2].

- *Anytime, anywhere communications.* Wireless networks provide the ability to have anytime, anywhere communications with minimal infrastructure. For example, services into remote regions, such as the Amazon, are often provided via wireless networks through technologies such as satellite [3].

- *In some cases, fewer anticompetitive concerns.* Generally many of the anticompetitive issues that arise with wireline service do not exist with wireless services. This is because the most popular wireless services, such as paging and mobile telephony, were generally deployed in a competitive environment and are provided in a competitive service market. Of course, there are exceptions to this, such as when governments impose spectrum caps on the amount of spectrum in which a single operator may have access to operate.

- *Less regulation may mean pricing advantage.* In many cases, wireless service operators have been able to escape having imposed on them the types of regulations that have burdened traditional wireline service providers, such as a universal service requirement or requiring specified accounting safeguards to be imposed. While wireless networks may face the imposition of increased regulation over the next decade, in cases where this has not yet occurred, wireless service providers may have an artificial price advantage over competing wireline service providers.

However, wireless services and networks are far from the perfect solution. There are many disadvantages with their use, including:

- *Propagation concerns.* Problems associated with the propagation characteristics of the radiocommunications spectrum, including rain fade, penetration into buildings, and the need for line of sight for clear communications often impact the availability and quality of a frequency band to a specific service.

- *Potential for interference.* Interference issues associated with nonconforming uses in a relevant spectrum band or from cofrequency services are always a risk. Wireline networks do not face this concern, as there should be no interference issues in almost all cases.

- *Difficulties in obtaining roof rights.* In order to obtain full coverage, extensive buildout is often required, especially with regard to terrestrial wireless services. This may be difficult to achieve because of the need for easements and access to rooftops and other rights of way in order to build towers, antennas, and other transmit/receive equipment.

- *Expense of regulatory fees.* Because the radiocommunications spectrum resource appears to be scarce, countries have begun charging more and more money for its use and setting fees based on market-based auctions. Hence, the regulatory fees associated with obtaining access to the radiocommunications spectrum may be onerous and negatively impact the ability of the service provider to meet its business plan [4].

- *In some cases, being reliant on the wireline network.* Many wireless systems are reliant, at least in part, on the wireline network. Accordingly, the success of wireless networks is often dependent on the extent of the cost to access this network and its availability in locations where it needs access.

Similarly, there are many benefits associated with the utilization of a wireline network. These include:

- *Reliable communications.* Wireline networks boast generally reliable communications services, without concerns about propagation characteristics, and they are less likely to face severe propagation delay problems. In most developed countries, for example, the availability of a wireline network is well above 99.95%, while wireless network availability is generally significantly below this percentage.

- *Eliminates interference concerns.* Because the communication travels via a wireline mechanism, the interference concerns associated with spectrum-based services is eliminated.

- *Network cost decrease as use increases.* Although an expensive network is required, once in place, the cost of the network decreases dramatically as usage increases.

In addition, in any evaluation of wireline versus wireless communications services, the disadvantages of the wireline network must also be considered. These disadvantages include:

- *The need for an imbedded infrastructure.* Service can only be required once an expensive imbedded infrastructure is put into place.

- *Fixed service only possible.* On a solely wireline network, service can only be provided to fixed points on an existing infrastructure.
- *Deployment may be slow.* Deployment of new services may be slow where the existing wireline infrastructure does not exist or is insufficient to support the relevant use. A good example of this is the

rolling out of high-speed Internet services by cable companies. In many cases, cable companies have had to rewire existing infrastructure to upgrade the infrastructure to support the bandwidth demands of this new service.

At the end of the day, the benefits and disadvantages of both types of services are evaluated by the service provider in determining the type of service they wish to provide and what type of technology they wish to use. In certain cases, such as in service to remote locations, wireless technology may be the only solution. However, if a service provider is able to rely in part on the existing wireline network, they may be able to decrease the cost of service provision by using a network made up of wireless and wireline components.

The key participants

As discussed, another driver in the consideration of the type of network to utilize is the point of view of the user. This section focuses on four key constituents, the types of uses they make of the spectrum, and some of the major issues that they are facing in the increasingly competitive search for spectrum.

The domestic government as user of the spectrum resource

In any country, one of the largest users of the radiocommunications spectrum resource is the government itself. Often, government users include the civilian defense ministries, scientific and educational uses, and public safety and distress uses. In most countries, however, the largest governmental user of the spectrum is the military. Like all assignments of the spectrum resource, the spectrum assigned to government uses is often under attack by advocates looking to use it for their own benefit. The next section discusses such efforts and also explores the issue of government self-regulation of spectrum use.

Self-regulation scenarios

In many countries, the government has no effective mechanism for controlling the efforts of government entities, including the military, from obtaining access to the radiocommunications spectrum, even for inefficient uses. Often, such use is purely a significantly less expensive alternative for these agencies in which to operate their communications. In these environments, the government entity may easily be able to access a desired frequency band, possibly even at the expense of other government users or private industry users. Further, governmental use that is unchecked may be inefficient and wasteful. A lack of a regulatory process for governmental use of the spectrum may mean that private operations are not provided access to portions of the radiocommunications spectrum that may be best utilized to provide widespread commercial applications.

In response, a few countries have either put into place or have proposed mechanisms for the self regulation of the government's use of the radiocommunications spectrum resource. A good example of such an approach is the bifurcation of the regulation of the spectrum resource in the United States. The FCC and the NTIA have split jurisdiction over the spectrum resource in the United States. Specifically, the FCC controls the use of the spectrum for commercial uses, while NTIA has that role for government use [5]. The FCC and NTIA coordinate continuously on such efforts, and both adopt and implement regulations that spectrum users under their jurisdiction must adhere. However, due to the pressure by commercial interests to free up spectrum that is currently used by the government, the U.S. Congress has recently been actively involved in ordering NTIA to identify government spectrum that can be freed up for commercial users.

A more novel approach that has recently been proposed is that contained in the recent U.K. Spectrum Review. Within that process, advocates have argued that government entities should be subject to economic forces just like other spectrum users [6]. For example, this proposed approach provides that government users should be allowed to trade their spectrum to the commercial sector and keep the funds earned from such trading as initiative to surrender unutilized or underutilized spectrum. This approach is very interesting but may end up handicapping the private sector in some instances by allowing the government to continue to hoard spectrum in the hope of being able to resell it later at a higher price.

Government use

Most of the spectrum that is utilized for nonmilitary government uses is for public safety and distress uses. Such uses can include police protection, safety-at-sea uses, aeronautical uses, and other similar uses. Many of these uses rely on spectrum-based services because of the nature of the communication. For example, aviation administrations utilize spectrum-based services for air-to-ground communications because it is impossible to use wireline facilities to complete the communication between air traffic control and airplanes.

The spectrum for such uses was generally assigned in the early days of telecommunications regulation and is generally seen as untouchable by commercial users of the spectrum. The reason the use of such spectrum is seen in this light is because of its public importance (i.e., it is politically difficult to argue that a commercial use of the spectrum, such as mobile telephony, is more important than air traffic control uses). However, this does not foreclose such efforts from occurring. To the contrary, if industry sets its sights on such spectrum, it may argue, for instance, that too much spectrum is assigned for such a use because a new technology has made more efficient operations possible or that the use is no longer valid because a new use has taken its place. Accordingly, the private sector could argue that it would be a more efficient use of the spectrum to allow a new use in these bands. Although such battles are often contentious, resource intensive, and time consuming, they sometimes result in a win for industry with the opening up of frequency bands for use by the private sector.

Spectrum that is used for nonpublic safety and distress or nonmilitary uses is often more likely to be sought for use by private industry because the political issues associated with such use are not likely to be as fierce. In some cases, governments may be willing to reassign such spectrum in exchange for private industry providing some of the functions that government has in the past. For example, some governments have allowed private industry to utilize spectrum traditionally assigned for education uses for private use, if they also provide educational services for free or for a nominal charge.

As discussed, one of the largest spectrum users in any country is the military. In most countries, the military is able to obtain and retain usage of key frequency bands because of its powerful and integral role in the government [7]. Accordingly, in examining most country's domestic frequency allocation and use tables, one would find that some of what industry would coin the most valuable or attractive portions of the radiocommunications spectrum resource assigned to the military.

Needless to say, as the telecommunications industry has grown, and as spectrum-based telecommunications systems have become more in demand by consumers, private industry has begun to challenge an increasingly large amount of this use [8]. Global industry believes in many cases that the military underutilizes the spectrum that has been assigned to it or uses it in a manner that is technically inefficient. In response, the military often argues that this is a flawed argument and works to entrench itself in the relevant frequency band. In other cases, the government flexes its muscle within other branches of the government to avoid even discussing this issue.

This conflict between the military and private industry is becoming even more common as the most attractive portions of the spectrum become more and more congested, and companies look to previously unusable bands for deployment of services.

However, it is a hard, uphill battle for industry to obtain access to military spectrum for several reasons. First, the military often operates under the cloak of confidentiality. Accordingly, in many cases, the military is able to block a wide inquiry into its use of a specified frequency band because of the confidentiality or security of its operations. Second, the military in most countries is extremely powerful politically. Accordingly, such battles often are fierce and reach into the highest rungs of the government for resolution. In this regard, only well-financed and politically well-positioned opponents stand a chance in such a battle. Third, in many cases the advocates of the proposed spectrum usage are vendors to the military. In this case, these advocates may not want to threaten their ability to retain the military as their customer.

In the late 1990s and early 2000s, prior to the events of September 11, 2001, it began to appear that the private sector was going to be successful in many countries in its efforts to obtain an increasingly large amount of spectrum traditionally assigned to the military. However, after these tragic events, and since the initiation of the global war on terrorism, the private sector's success in its efforts is less than certain, especially in the United States and European Union member states [7].

Telecommunications service providers and broadcasters

One of the largest spectrum constituents is the telecommunications service providers and broadcasters. Telecommunications service providers are the operators of telecommunications networks, such as Telefonica de Espana in Spain, AT&T Wireless in the United States, and Korea Telecom in South Korea. These service providers may provide a wide range of services or a single telecommunications service and may utilize the resources of network operators, such as PanAmSat, to provide services. Broadcasters, on the other hand, may include entities such as the powerful U.S. networks for NBC, ABC, or CBS, or the United Kingdom's BBC, or more local broadcasters.

Telecommunications service providers

Providers of wireless telecommunications services have become increasingly aggressive in the market as they recognize that "radio spectrum is the key 'asphalt' for the latest generation of the Information Highway, wireless Internet" [9]. In addition, many service providers today operate in more than one country.

In order to operate their telecommunications systems, these service providers must obtain authorizations and assignments from individual governments for each proposed use, which also specifies the frequency band, geographical area of service, and any technical rules with which the provider must comply. Accordingly, a company that wants to provide mobile telephony in Paris and London must obtain individual regulatory authorizations from the relevant regulator in France and the relevant regulator in the United Kingdom for such service. Further, if that same operator wants to also operate a wireless cable service in London, it must obtain a separate authorization for that operation from the United Kingdom regulator that would have the authority to allow such use.

In today's telecommunications world, usage of the radiocommunications spectrum resource has become increasingly important to telecommunications service providers as a method of providing services both directly (as in mobile telephony) and indirectly (as an adjunct to existing services, such as providing a last-mile connection to the home through point-to-point microwave services). Accordingly, many companies hold multiple authorizations for multiple uses in the same country or even geographical area.

In some cases, private industry may be closely aligned in obtaining spectrum. For example, when seeking an allocation of an individual frequency band to a specific service or identifying a frequency band for a specific use, several telecommunications service providers who support such a cause may band together in support into either a formal or informal association. As discussed in Chapter 8, such joint action often adds credence to the advocate's efforts and provides additional political pressure on the regulator to act in a specific manner.

There are also cases where industry is diametrically opposed. For example, once a frequency band is allocated to a specific service and identified for a set use, companies that were formerly allies may now be seeking assignments of the same spectrum. In this situation, a fierce battle in the authorization and assignment process may occur. For example, although many wireless mobile service providers worked together jointly at WRC

2000 in order to obtain access to a common frequency band for 3G services, they were often fierce competitors as they bid on regulatory authorizations to provide 3G services around the world.

Further, industry is often in an adversarial position toward one another when an advocate of a new use seeks to utilize spectrum that is already being used by another service provider for an existing service. In such cases, the incumbent user will often fight a fierce battle to preserve its ability to use the spectrum where it is currently operating. A good example of such a battle was the successful effort of the MMDS community in the United States to keep the 3G service providers from utilizing the 2.5-GHz band for their services.

In some cases, the incumbent user or new entrant may be part of a former government monopoly and have continuing, although indirect, ties to the government, which provides it with certain political advantages in a fight. Good examples of this are NTT in Japan, which was part of the Japanese Ministry of Posts and Telecommunications, and British Telecommunications, which was formerly a part of the agency that was also the spectrum manager in the United Kingdom. These entities may be looked at more favorably by the regulator than an unknown entity because of the past relationship.

If such a battle looks like a loss, the incumbent user may compromise and seek relocation (and corresponding payment from the new user) to another frequency band where it can also operate. This was the case in the United States when Teledesic sought interference-free operation in the 18-GHz band from point-to-multipoint operators. Ultimately, the parties, working with the government, formed a consensus solution that satisfied the needs of all parties. This resulted in the relocation of the point-to-multipoint operators to a different but acceptable frequency band. In many cases, this type of compromise requires the new use to pay for the relocation of the existing use to a different frequency band.

In more liberalized telecommunications markets, success on the part of a telecommunications operator in obtaining new spectrum or retaining old spectrum for use often depends on many factors. These factors may include, among others:

- The political power of the advocate;
- The amount of resources the advocate is willing to expend to fight to utilize the relevant frequency band;

- The public-interest benefits of the service that can be demonstrated by the service provider to the government;
- The types of services to be provided and consumer demand for such services;
- What sort of commitments the service provider is willing to make in order to be able to offer its proposed services;
- The lasting power of the participant, as spectrum battles quite often take years to resolve.

However, the factors used to evaluate such success in less competitive or closed telecommunications markets are less certain. In such cases, the political will of the government is often key to any success.

Broadcasters

Another important category of spectrum users is the television and radiocommunications broadcasters. National broadcasters, especially, such as the United Kingdom's BBC, hold access to vast spectrum assets and have significant political clout because of their reach into the general population. Often, in large part because of their public-service mandate and far reach, broadcasters are considered a specialized service and are not regulated as part of the rest of the radiocommunications spectrum. Accordingly, many governments have established separate agencies to regulate the broadcasters. A good example of this is in Nigeria, where the government is setting up three different spectrum-management agencies: one for government spectrum, one for broadcasters, and one for the nonbroadcast private sector. By arranging a spectrum bureaucracy in such a manner, the government may be able to further protect access to the broadcast spectrum by other members of the private sector.

Telecommunications equipment manufacturers

Another key constituent group with regard to the radiocommunications spectrum is telecommunications equipment manufacturers. For the purposes of this book, telecommunications equipment manufacturers refer to manufacturers of backbone equipment (e.g., large antennas, switches, and satellites) as well as the manufacturers of consumer end products (e.g., mobile telephony handsets). Examples of some equipment manufacturers who are very active and powerful in the area of radiocommunications devices include Nokia from Finland, Nortel from Canada, Alcatel from France, Samsung from South Korea, and the U.S. manufacturers Lucent and Hughes.

The prime motivator behind the intense activities of the equipment manufacturers in the field of radiocommunications is sales. Quite simply, they want to ensure that their customers, both the telecommunications service providers and the end users, have access to the portions of the radiocommunications spectrum that their devices can operate in and that this spectrum is allocated for use by the relevant services and allows technically their operation. Accordingly, both domestically and internationally, the equipment manufacturers are active in ensuring the availability of spectrum for the uses that they are most interested in manufacturing equipment in which to operate.

A good example of such activities by equipment manufacturers is the efforts made by the 3G equipment manufacturers at WRC 2000, at its preparatory meetings, and in accompanying domestic proceedings. In this regard, equipment manufacturers such as Nokia and Motorola actively sought out sufficient spectrum for the operation of 3G services in the frequency bands in which they felt it was optimal for their equipment to operate. In many cases, the equipment manufacturers, more so than even the telecommunications service providers, led these efforts because of the direct financial impact on these manufacturers. Another reason for this is the lag of the technical market. This results in a dynamic whereby equipment manufacturers are often ahead of the service providers in planning for new services. Accordingly, before manufacturing the relevant equipment, these entities will look for certainty.

An interesting trend that has been occurring in recent years is the growing desire on the part of equipment manufacturers for global allocations of a single frequency band for an individual service and identification for use of an individual frequency band for specified use. This trend has been most prominent in Europe and Asia, where manufacturers feel that set standards make the manufacturing process easier to work with, as a single piece of equipment will work anywhere. The United States has consistently pushed against such an approach, believing instead that the marketplace should be the ultimate arbiter of technology. It is likely that as telecommunications markets become increasingly global and as uses continue their trend towards transborder usage, countries will work in a more coordinated effort in designating or identifying spectrum for specific applications. Failure to do so may result in a patchwork of equipment devices that will not work in all countries, ultimately leaving the consumer unconnected.

In all cases, however, it must be remembered that new technology is not a spontaneous occurrence. With regard to radiocommunications equipment especially, research and development is based on both market demands and the regulatory climate. Accordingly, there must be not only a need for such equipment, but the regulatory regime governing the proposed technology must allow for it or be changed to allow for it. This is somewhat different than what happens in other high-technology fields, where often the best technologies are created without transparency and then released without advance notice.

Accordingly, manufacturers are often largely constrained by the amount of regulation to which telecommunications technology is subject. In some countries, such as the United States, freedom of technology by service providers is authorized and a desired end result. The U.S. philosophy is to let the market decide what technologies will be utilized in a particular frequency band to offer the desired telecommunications service. However, as discussed, many countries, including those of the European Union and Japan, feel that technologies should be dictated by governments and adhered to. What these countries fail to recognize, however, is that by picking technical winners and losers, they are inhibiting technical innovation.

Consumers

In general, all consumers have the same general goal: to obtain reasonably priced high-quality telecommunications services. However, divergent interests exist between the different groups of consumers. In this regard, consumers can be broken into two different groups:

- *High-end users:* large and medium enterprise consumers, such as multinational corporations or hotel chains, and high-profile users (such as celebrities or corporate executives);
- *General consumer users:* residential or small business consumers.

High-end users are generally looking for the most reliable means of transmitting their communications information to all their operations at the most reasonable rates. Of course, different types of high-end users may also have additional needs. For example, for a large global bank like Citibank, the security of the transmission may be of increased importance, whereas an airline, like United, may require service-level guarantees of 100% reliability because of safety concerns. General consumer users, however, are often willing to accept lower quality services in order to obtain better prices.

The interests of these groups directly impacts what frequency band telecommunications service providers may be willing to operate in and what accompanying technical rules may be acceptable. For example, a service provider that is primarily looking to serve residential services may be willing to operate in a frequency band with a slight potential for interference. However, a service provider who is looking to provide the highest quality of service possible to demanding multinational companies may not be willing to operate in the same frequency band or with the same technical constraints on operation.

In addition, the ability for both high-end and general consumer users to utilize their communications equipment internationally is also important. Accordingly, a substantial amount of time and effort has been invested in the ITU and other forums in establishing a regime that allows wireless communications devices, such as mobile telephony handsets, to be freely brought into other countries. Under the agreements that have been reached on this issue, such as the ITU's Memorandum of Understanding on Global Mobile Personal Communications Devices, companies that abide by the technical standards established in the agreement are able to produce equipment that is freely transportable by consumers into multiple countries without obtaining additional nationalistic-type approval of the equipment.

General consumer users are actively involved in the spectrum arena battles generally only when they need a service and are trying to preserve an existing service's availability or trying to influence a proposed change that will directly impact them. In such cases, consumers may work on their own, with other consumers, or with other participants to obtain a satisfactory resolution.

Factors impacting the use of the spectrum resource

Now that a firm understanding of the major participants in the spectrum arena has been established, it is helpful to understand the significant primarily nontechnical factors that directly impact the use of the spectrum resource. These factors include:

- The government regulator and the accompanying regulatory regime;
- Market demand for the service;
- Amount of spectrum available for the same or similar use;
- The cost of obtaining access to the spectrum (including regulatory fees, research and development, and equipment) and the impact on the business case;
- The ability and cost to use terrestrial landline networks for the same service;
- The ability to obtain access to rights of way for network buildout.

Each of these factors is addressed in more detail in the following sections.

The government regulator and the accompanying regulatory regime

No discussion of the radiocommunications spectrum would be complete without focusing on the domestic government in its role of regulator of the spectrum. Each constituent group is directly dependent on the regulator or regulators of the radiocommunications spectrum to allocate and assign spectrum. While the spectrum allocation and assignment process will be the subject of more in-depth discussion in Chapter 5, it is important to have a broad understanding of the role of the regulator and the governing regulatory regime at this point.

The role of each domestic regulator of the spectrum resource generally includes:[2]

1. Allocating individual frequency bands of the radiocommunications spectrum domestically to specific services (in accordance with international obligations);
2. Authorizing specific uses of the radiocommunications spectrum within individual frequency bands;

---

2. In this context, when the term domestic regulator is used, it is in reference to the reulator or regulators governing both public and governmental use of the radiocommunications spectrum resource.

3. Assigning the radiocommunications spectrum resource to individual users or operators for a limited term and under specified terms and conditions.

Accordingly, it is the domestic regulator that ultimately determines what use will be made of a specified frequency band, who will be able to operate and use that frequency band, and what limits will be placed on operations. As discussed subsequently, the processes that the domestic regulator(s) utilize to make each of these determinations directly impacts the availability of spectrum for a particular use, the ability to utilize that spectrum for that use by an individual operator or user, and the cost to obtain access to that spectrum. In many cases, as outlined in subsequent portions of this book, in order to obtain access to a specific portion of the spectrum resource, operators and other users will launch extremely resource-intensive efforts to gain or retain access to the spectrum resource for their specified usage. Often, such efforts are the equivalent to outright battles, which are also known as spectrum wars. The efforts of the 3G providers to obtain access to additional spectrum at WRC 2000 and in domestic arenas since then and the efforts of Teledesic to obtain spectrum both globally and on individual domestic basis for its NGSO satellite system have been among the most notable of these battles in the recent past.

Of course, the process that is utilized for each of these responsibilities is dependent on the specific regulatory regime. In a closed market or one with limited competition, it is unlikely that the private sector will have much of a role in establishing the rules governing the allocation, assignment, and designation or identification of the spectrum resource. However, in more competitive markets, and especially in countries that have adopted the regulatory principles encased in the World Trade Organization's Basic Agreement on Trade in Telecommunications Services (WTO Agreement), it is extremely likely that private industry will play a direct role in developing each of these issues.

The WTO Agreement is the cornerstone treaty on the free trade of telecommunications services. The WTO Agreement sets out a framework for market liberalization of telecommunications services, which includes:

- Market access and national treatment;
- Foreign investment;
- An international dispute settlement mechanism for failure to meet commitments.

In addition, it addresses the imposition of certain regulatory principles, including [10]:

- Transparency in the process;
- The creation of an independent regulator;
- The implementation of competitive safeguards;
- Fair and nondiscriminatory interconnection.

Accordingly, a firm understanding of each government's regulator and regulatory regime is imperative for the constituent to understand how to best obtain its goals.

Market demand for the service

In any evaluation of the use of the radiocommunications spectrum, it is imperative that an understanding of the market demand for the service be evaluated by taking into account the actual cost of the service to the end user and the technical characteristics of the service. An inability to understand this dynamic may lead to failure on the part of the service provider from an economic perspective.

Unfortunately, it is often difficult to obtain a good understanding of market demand for a new, unproven wireless telecommunications service. An excellent example of where such market demand was misunderstood involved the Iridium satellite system. The initial concept for the Iridium satellite system was the deployment of a 66-NGSO satellite constellation to provide mobile services to high-end consumers. Use of the satellite system would cost approximately $10 per minute for phone service, and the handset cost well over $1,000. To minimize the phone service cost, however, the handset was multiband, which allowed it to switch to terrestrial mobile service when such service existed. This would lower the price of the service in such cases to be comparable with existing mobile telephony service.

Unfortunately, Iridium overestimated its consumer attractiveness. First, consumers found the service expensive to use, especially because its deployment began around the time that mobile telephone prices first began to drop dramatically. Second, Iridium overestimated the demand for "anywhere" type of phone service. At the end of the day, the number of customers who needed to be reached anywhere in the world was dramatically less than estimated. In addition, early usage of the Iridium service demonstrated a less than perfect system, with early users facing technical problems. Further, the handsets that were created were large and cumbersome. Finally, and perhaps most importantly, failure can be traced to the timing of the release of the Iridium system. By the time the Iridium system was ready for service, mobile telecommunications service through nonsatellite means was virtually ubiquitous through roaming agreements and national buildout. All of these factors led to the bankruptcy of Iridium. However, through scaling back its service plan and revising its business case, today Iridium has emerged from bankruptcy and is currently providing service.

Accordingly, to avoid similar results, many companies expend substantial resources evaluating the market demand of the proposed service. Of course, estimating the demand of a new service is always difficult, especially when you are depending on global customers. Therefore, with new and innovative wireless services, there is often an inherent risk in such deployment.

Amount of spectrum available for the same or similar use

Another key consideration is the amount of spectrum that is available for the same or similar uses. This consideration ties in directly with correctly understanding the market demand for service. In general, it is important to understand whether there will be a tremendous influx of the same or similar service providers with which the provider will have to compete. Because of the large geographic reach of spectrum-based services, it is generally more cost-efficient to have a broader service area, in terms of population coverage, and no competitors or only one competitor. However, whether such limited competition serves consumers is questionable—because they will only have limited choice in service providers. Service providers, in response to such an approach, argue that unlimited or increased competition will only result in increased prices to consumers because the providers will have substantially less market share with which to finance their telecommunications system.

Another spectrum consideration is the value of the spectrum to the applicant or user. For example, if there are only two assignments available for a specific use and more applicants, then the amount of resources (e.g., money or time) that the proposed user will expend increases substantially. This is dramatically different if there are more assignments or a similar number of assignments available than there are interested users.

The costs of obtaining access to the spectrum and the impact on the business case

Probably one of the largest drivers of the use of the radiocommunications spectrum resource is overall cost of access. These costs can include:

- Regulatory and other fees, which include any licensing fees, taxes, or other regulatory fees that are required to obtain and retain use of the spectrum (in some cases, instead of working directly through the regulatory process, the service provider can obtain spectrum through the secondary market that is beginning to develop for access to the spectrum resource);
- Costs involved in any regulatory actions that are required to ensure that use of the planned frequency band is available for the relevant use;
- The costs associated with research and development of the telecommunications service and accompanying equipment;
- The costs of obtaining easements and other rights of way in order to build out infrastructure;
- The cost of equipment to provide the telecommunications service;
- The cost of consumer equipment.

Each of these costs is critical in developing the telecommunications service provider's business case. Failure to account for such costs in a realistic manner can result in overly optimistic rates of return. This is likely what happened in the European 3G bid auctions, after licensees paid substantially larger than expected auction fees for the 3G spectrum and then faced huge financial pressures during the buildout of their systems, in some cases calling into question their continued viability.

The availability of terrestrial wireline infrastructure

In some cases, it may not always be cost-effective to provide all of a wireless service on a wireless basis or it may be technically necessary to operate a wireless and wireline network together. For example, many mobile telephony networks use wireline networks to carry the traffic from some of their cell sites to their switching station. Accordingly, any analysis must include an examination of the availability of the terrestrial wireline infrastructure for use and the costs, benefits, and disadvantages of utilizing such a network with the wireless network. In many cases, as discussed, such an analysis may result in a determination that a combined wireline and wireless network would be the most efficient solution to provide the proposed use.

Conclusion

This chapter has provided an overview of the advantages and disadvantages of wireless telecommunications services and networks (in contrast to wireline services and networks), an introduction to the interests involved in spectrum battles, and an exploration of the key considerations involved in deploying a wireless telecommunications network and providing service. This provides a firm basis for the exploration of the domestic and international processes and structures that govern the radiocommunications spectrum, which are discussed in Chapters 4, 5, and 6.

Endnotes

[1] Geier, Jim, "Benefits of Wireless Networks," August 2000, available at http://www.wireless-nets.com/articles/whitepaper_wireless_benefits.htm.

[2] Kant, Eric, "Wireless GIS Solutions Aim WTC Rescue Efforts," 2002, available at http://www.esri.com/news/arcuser/0102/wtc1of3.html.

[3] Intelsat in Africa, available at http://www.intelsat.com/news/mediakit/press_ex/intelsatsfrica.asp.

[4] However, over time, some governments are lifting spectrum caps; see Mosquera, Mary, "Feds Lift Spectrum Caps on Wireless Providers," *Internet Week,* November 6, 2001, available at http://www.internetwk.com/story/INW20011109S001; see also Glasner, Joanna, "How Spectrum Caps Turned Sour," *Wired,* March 5, 2002, available at http://www.wired.com/news/wireless/0,1382,50782,00.html.

[5] Roosa, Paul Jr., "Who Regulates the Spectrum," *Federal Spectrum Management: A Guide to NTIA Process,* NTIA Special Publication 91-25, August 1992, available at http://www.ntia.doc.gov/osmhome/roosa4.html.

[6] http://www.spectrumreview.radio.gov.uk.

[7] Rucker, Teri, "The Military's Spectrum Pitch: Our Calls Must Go Through," *Technology Daily,* April 23, 2002, available at http://www.govexec.com/dailyfed/0402/042302td1.htm.

[8] See Mosquera, Mary, "Spectrum Search Puts Military in the Hot Seat," *Internet Week*, July 25, 2001, available at http://www.teledotcom.com/article/TEL20010725S0002.

[9] Industry questions and answers at http://www.wow-com.com/indstry/policy/congaffairs/articles.cfm?ID=437, June 7, 2001.

[10] See "Reference Paper to Basic Agreement on Telecommunications," available at http://www.wto.org.

4

The Regulatory Regime Governing Spectrum

Why is the radiocommunications spectrum resource regulated?

In order to ensure that both coordination and use of the radiocommunications spectrum runs smoothly and efficiently, governments actively regulate the spectrum resource through domestic and international processes. Almost all governments and users of the spectrum resource believe that some form of regulation over the radiocommunications spectrum is necessary. The amount and type of regulation that is necessary, however, is often disputed among different interests.

The primary reason that the regulation of the radiocommunications spectrum is so important is that it is a scarce resource. It became readily apparent in the early years of radiocommunications to almost all governments that this resource should be managed in a manner that ensures its efficient use and guards against harmful interference among competing uses. Similarly, because the spectrum resource, unlike the wireline network, does not recognize or respect the arbitrary boundaries of individual nations, it was recognized early that *global* coordination of use of this resource was necessary. Such coordination is also important to ensure that interconnection is possible between radiocommunications devices of different countries and manufacturers.

Each of these concerns provided an impetus for countries to forfeit some of their sovereignty in exchange for having a coordinated utilization of the global spectrum resource. Accordingly, an international regulatory framework for spectrum has been created, with the ITU having primary regulatory responsibility. Several key areas that the international regulatory process addresses include:

- The international allocation of radiocommunications spectrum;
- The creation and implementation of radiocommunications spectrum allotment plans;
- The setting of international technical standards for use of specific frequency bands;
- Bilateral and multilateral radiocommunications spectrum coordination.

However, international regulation has not eliminated the rights and sovereignty of individual nations to utilize the spectrum in the manner in which they see appropriate, as long as it does not result in harmful interference to the services of other countries operating in compliance with the ITU Radio Regulations and International Table of Frequency Allocations. Key areas that domestic regulatory processes address include:

- The domestic allocation of radiocommunications spectrum;
- The domestic use of radiocommunications spectrum;
- The domestic assignment of radiocommunications spectrum;
- National inputs for international radiocommunications spectrum issues;
- Coordination of competing uses both domestically and with neighboring countries.

In the international regulatory arena, individual companies, commercial interests, and other interested parties often aggressively seek to achieve their nationalistic goals. They do so through both the established international negotiation process as well as through political means. Depending on the importance of a particular issue to a government, political pressure may be placed on other governments to obtain support for that position. In addition, there may be compromises reached on other issues of concern, so that support can be given by one country to another on the particular spectrum-related issue.

Commercial service providers, government users, and other users of the spectrum also work aggressively in each of the regulatory arenas to obtain their own goals for use of the spectrum resource. As discussed in Chapter 3, commercial service providers and equipment manufacturers often operate in multiple countries. Accordingly, these providers may be active in numerous countries' individual regulatory arenas, as well as in the international arenas. Each interest is likely to actively lobby the decision makers in the relevant market where they operate in order to obtain support for their positions. In the case of private industry, support is generally obtained by explaining the economics and public-interest benefits of their proposals. Government users make public-interest arguments, but also rely on arguments based on national security and other governmental goals.

Accordingly, the domestic and international regulatory arenas are major battlefields for different interests to meet and advocate their own use of the spectrum, while arguing against the use of the same frequency bands by others or guarding against interference from uses of the spectrum operating in adjacent frequency bands. Further, once the allocation and use of the spectrum is determined, there are equally fierce battles over how to best assign the spectrum to individual operators in individual countries. Most recently, a third battlefield has been forming. This is a secondary market for radiocommunications authorizations and assignments that are held by companies that may not have sufficient resources to build out their networks. A good example of this is the recent Nextwave settlement reached with the FCC [1]. In this case, the bankruptcy courts ruled that the FCC does not have the authority to reauction a bankrupt licensee's previously purchased spectrum [2].

This chapter discusses the goal of spectrum regulation and provides an overview of the international and domestic regulatory process governing the radio spectrum resource. This chapter also focuses in detail the international process governing spectrum allocation and regulation. Chapters 5 and 6 are the second part of the regulatory discussion and focus on the domestic regulatory process for the allocation, use, and assignment of spectrum. In addition, Chapter 8 focuses on the secondary market that is being created for spectrum assignments in many domestic markets. Each of these areas is integral to understanding the underpinnings of the many battles over the radio spectrum that have occurred and will be faced in the future.

The goals of spectrum regulation

The goals of most domestic regulators and international bodies in the regulation of the spectrum resource are fourfold. Specifically, these goals are:

- To make certain the efficient use of the radiocommunications spectrum;
- To properly manage the scarce radiocommunications spectrum resource to ensure its use is maximized;
- To coordinate radiocommunications spectrum uses to guard against the potential for harmful interference;
- To ensure that radiocommunications systems of different countries can interconnect with one another.

In allocating and assigning the spectrum resource, these four goals serve as the prime motivators for most governmental actions. These goals are often impacted by other concerns or subsidiary goals, because regulation of the spectrum does not occur in a vacuum. Subsidiary goals may also influence the regulatory decisions of government bodies in the spectrum arena. For example, in times of war, a government may be less likely to reallocate military spectrum to the private sector than it would be in peaceful times. In addition, if political goals are outstanding between governments, such as over economic crises, schisms will also be recognized in other forums, such as that governing the radiocommunications resource.

Some examples of subsidiary goals include:

- National security;
- Privacy;
- Security;
- Public safety;
- Foreign policy;

- Military;
- Welfare.

Other subsidiary goals sought by governments often depend on the competitive status of the telecommunications market. For example, in a telecommunications market that is fully open to competition, a key governmental goal may be to create spectrum allocations or adopt accompanying technical rules that are responsive to the competitive marketplace. However, a government in a noncompetitive market, such as China, would be substantially less likely to pursue such a goal.

In addition, regulators must also consider private-sector concerns, where the main motivation in spectrum allocation and assignment will often be economic. These private-sector members may include service providers, equipment manufacturers, and users. The ability of the private sector to influence the process is largely dependent on the competitive status of an individual market and the openness of the process to the private sector. In fairly competitive and open telecommunications markets, like the United States, Brazil, and the European Union member states, private-sector goals may have a significant impact on the actions the government takes both domestically and internationally with regard to spectrum regulation.

Overall, each action or position a government takes on spectrum issues will be impacted by how its subsidiary goals impact the four main policy goals for spectrum regulation and the position of the private sector. Accordingly, the ultimate decision of a government on an international position or a domestic action will often be the result of a very carefully crafted balance of interests and goals. This was exactly the case of the U.S. position at WRC 2000 with regard to the identification of spectrum for 3G services. As discussed earlier, the U.S. position was to encourage each country to choose between the 1.8-GHz and the 2.5-GHz bands, if any spectrum was to be used for 3G services.

In this situation, the United States had to balance many goals including:

- Ensuring the most efficient use of the spectrum resource;
- Not prejudging its own domestic process on the assignment of spectrum to 3G services;

- Enabling the ability to deploy new technologies that are responsive to the public.

In order to accommodate these and other goals, the United States adopted a very flexible approach that would allow each ITU member state to choose whether to assign spectrum for 3G services and in what frequency band [3].

The governing regulatory bodies

The spectrum regulatory regime is essentially bifurcated with the ITU (with input from regional bodies, member states, and private-sector members or participants) having the lead role internationally and individual domestic regulators being responsible for domestic spectrum regulation. The ITU and domestic regulators have a symbiotic and circular relationship with each party reliant on the other for the spectrum resource to be utilized in a manner that ensures their goals of spectrum regulation are met.

Because the international allocation of spectrum and the adoption of corresponding technical rules directly impact what use domestic regulatory authorities can put their spectrum to, ITU-R activities, most notably the WRCs, are seen as the very important first step in the process for use of spectrum. Accordingly, almost all countries tend to participate in the WRC process in order to input their parochial interests into the process. The international arena is often contentious, and countries often undergo elaborate domestic processes before important international meetings to determine their position on key issues of interest to them. However, once the international battles are resolved, that is not the end. The battle is then simply removed once again to the individual domestic forums, where equally fierce battles take hold on how to implement the decisions of the ITU domestically.

Many regional bodies actively feed into the ITU process and heavily shape the outcome of spectrum issues at WRCs. These bodies include:

- *Conference of European Posts and Telecommunications Administrations* (CEPT), which is a standard-setting body made up of regulators from around the European and African regions;

- *Inter-American Telecommunications Conference* (CITEL), which is an arm of the Organization of American States, wherein governments and private-sectors members focus on telecommunications issues that are of interest to the region;
- *Asia-Pacific Telecomunity* (APT), which is a body that represents governments in the Asia-Pacific region on telecommunications issues and works towards finding consensus in the region.

The regional bodies generally focus on forming consensus positions within their region that will be jointly advocated at upcoming meetings, culminating at the relevant WRC. To date, CEPT has been the most successful regional body at pursuing this approach. This was readily apparent at WRC 2000, where CEPT was instrumental in ensuring that spectrum was identified for 3G use. In fact, in part, the strength of CEPT forced other countries, including the United States, to find a compromise position on the issue of the identification of spectrum for 3G services, despite the U.S. government's initial hesitancy to support any such identification.

This increased coherence among the CEPT countries is beginning to be experienced among the other regional bodies as well. It is anticipated that at WRC 2003 this will become more apparent, especially as equipment manufacturers and some global service providers continue to believe that global coordination of the use of the spectrum is of increasing importance, such as was the case in the identification of spectrum for 3G use.

The international regulatory process

As discussed, the international regulatory arena governing radio spectrum issues is dominated by the ITU. The ITU is comprised of member states that take action at WRCs on the allocation of spectrum. This process takes place through a formal treaty negotiation that occurs once every 2 to 3 years and last for about 4 weeks [4].

The WRC meetings are often quite contentious. Discussions and decisions are as often politically motivated as they are technical. While member states strive toward consensus decisions, votes may occur under extreme circumstances. More likely than a vote, however, a chair of a meeting may take an informal vote in order to force a consensus decision to be agreed to. Further, if a member state chooses, it can take a reservation to any decision adopted by the WRC if it does not support such a decision. This means that the reserving country does not have to abide by the decision of the conference, as long as such action does not cause harmful interference to the stations of other countries operating in conformance with the ITU Radio Regulations.

To date, the WRC process has been successful. For the most part, countries abide by the International Table of Allocations and the ITU Radio Regulations, even though the ITU has no real enforcement mechanism at its disposal. Most countries have found it in their interest to comply because of international pressure and the need for global coordination of the spectrum resource to ensure that the radiocommunications devices of all countries can operate without harmful interference occurring.

An overview of the ITU and the Radiocommunications Sector

As discussed, the ITU is a treaty-making arm of the United Nations with certain responsibilities governing telecommunications regulation. Among these are the mandate of Article 44 of the ITU Constitution, which provides for the exercise of care and consideration in the use of the spectrum [5].

The ITU is divided into three different sectors: the Telecommunications Sector (T Sector), the Development Sector (D Sector), and the Radiocommunications Sector (R Sector). The R Sector of the ITU has the most direct responsibility over the radiocommunications spectrum resource, although there is some overlap with the other sectors [6]. The main responsibilities of the R Sector, as provided for in Articles 1 and 12 of the ITU's Constitution are [7]:

- To ensure the rational, efficient, economical, and equitable use of the radio frequency spectrum by all radiocommunications services;
- To carry out studies without limit of frequency range on radiocommunications matters and to adopt suitable recommendations.

The procedures of the R Sector provide confidence-building measures as to the use of spectrum to both governments and private-sector users. First, certain guidelines governing the use of the spectrum require international agreement to change. This provides service providers and equipment manufacturers with a certain amount of design and operational certainty. Second, at least as far as harmful interference occurring from neighboring countries to radiocommunications services, time-tested methods can avoid and solve such circumstances.

The R Sector carries out its responsibilities in large part through the issuance of the ITU Radio Regulations and the allocation of spectrum in the ITU Table of Frequency Allocations. The ITU Radio Regulations and the Table of Frequency Allocations (which is part of the Radio Regulations) can only be changed by the WRCs. WRCs can also establish assignment or allotment plans to secure radio transmission or reception within individual national boundaries (although individual member states are responsible for the assignment of spectrum to individual users through the licensing process). Where such plans are not utilized, international coordination procedures are put into place to ensure that harmful interference does not exist among countries when use is made of the subject frequency band.

WRCs are generally decided on the basis of consensus decisions. However, when votes are taken, it is important to remember that the ITU works on the premise that each member state has one vote. Accordingly, it is important for advocates of a particular position gather sufficient support among member states so that they have enough support for their position to be carried by the majority of member states at the WRC.

Overview of the ITU R Sector

As defined earlier, the R Sector is a largely bureaucratic organization comprised of several bodies (see Figure 4.1). One key body of the R Sector is the *Radio Regulation Board* (RRB). The RRB approves the rules of procedure that are used in the application of the Radio Regulations to register frequency assignments. The RRB considers any matter that cannot be resolved through the application of the Rules of Procedure. The RRB also has the important role of assisting in mediating disputes between member states on issues of harmful interference [8].

The Radiocommunications Bureau is the administrative arm of the R Sector. Two of its key responsibilities are [9]:

- Administering the Master International Frequency Register;
- Assisting in the resolution of cases of harmful interference among members.

The R Sector also has many different study groups operating under its auspices. These study groups are broken out as follows [10]:

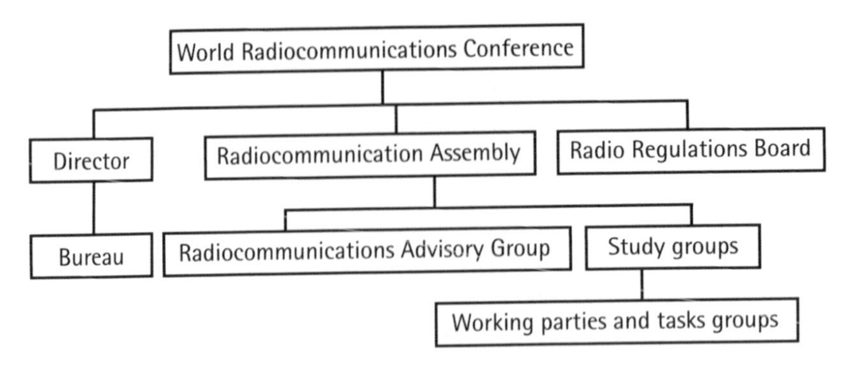

Figure 4.1 Overview of the ITU R Sector. In addition, the Radiocommunications Bureau provides administrative support at R Sector meetings, including at WRCs. (*After:* [11].)

- Study group 1: spectrum management;
- Study group 3: radiowave propagation;
- Study group 4: FSS;
- Study group 7: science services;
- Study group 8: mobile, radiodetermination, and amateur and related satellite services;
- Study group 9: FS;
- Study group 10: sound broadcasting;
- Study group 11: television broadcasting;
- Study groups 2, 5, and 6 have been eliminated.

Within the confines of the study group process, technical and other experts from across the globe study technical issues of relevance to the R Sector. The recommendations of these study groups are key to the outcome of issues addressed at WRC 2000. For example, the outcome of study group 8 was the guiding principle for the amount of spectrum that would be identified to 3G services at WRC 2000.[1]

1. There are also other organizations within the R Sector, such as rappatuer functions, which are beyond the confines of this book.

The impact of regional organizations on spectrum regulation

In recent years, the role and importance of international organizations in the global radiocommunications spectrum process has increased dramatically. These organizations, through their member states and, where appropriate, with private-sector participation, participate in forming consensus positions for WRCs and coordinating positions on spectrum issues for other key meetings.

Specifically, the key organizations that are often instrumental in the spectrum world include:

- *CEPT:* CEPT promotes cooperation between member administrations and bodies responsible for telecommunications policy and regulation. Its activities include spectrum management. CEPT has become a force to be reckoned with by many other countries. Because of the large block of countries that are CEPT members, when CEPT presents a position at an ITU meeting, including WRCs, it is clear that a substantial number of countries support it (even if not unanimously). CEPT has more than 40 member states from Western Europe, Eastern Europe, and Africa. Despite the strength of CEPT, there are times when CEPT member countries will actively work against CEPT positions, although these are rarely, if ever, EU member states.

- *CITEL:* CITEL is the principal advisory body to the Organization of American States on telecommunications issues. CITEL's main objectives are to promote the development of telecommunications in the Americas in order to contribute to the overall development of the region. CITEL has many member states, including the United States, Brazil, Argentina, Mexico, and Canada. In recent years, CITEL has begun to become more cohesive, at least among the more developed and active member states of the organization. This has largely been in response to the increasing importance of CEPT.

- *APT:* APT is the principal body for the Asia-Pacific region in terms of determining positions at the WRC. Key active members include Japan, Australia, and South Korea. Its membership covers 32 member states. APT first started its regional coordination work on radiocommunications in 1996, in preparation for WRC 1997. Although

APT is a fairly new organization, its importance in the international spectrum allocation process is well recognized.

The Middle East and the African states are also becoming more cohesive at WRCs outside of the confines of CEPT, and they are beginning to form active and unified regional organizations with the goal of coordinating positions at WRCs and other relevant international meetings. Because of the principle of *one country, one vote* at the ITU, the large numbers of countries in these areas have the potential to steer the direction of a decision. This has happened at past WRCs where whole conferences have been forced to pay heed to the needs of developing countries in their compromise solutions. Accordingly, ITU member states that are actively searching for spectrum will often expend substantial resources negotiating with these countries in an effort to obtain support for their position. The lesser developed countries, in large part because of their sheer voting power, are able to directly influence the outcome of a WRC.

Traditionally, regional bodies played little or no role at WRCs. However, as regulation has grown, and especially in response to the growing cohesion among European regulators due to the increased importance of the EU, increased use of these organizations has been made to influence the outcome of ITU conferences, including the WRCS. For example, CEPT currently has 47 member states. If a member of CEPT is able to count on the support of a significant portion of these member states on a CEPT position they are sponsoring, they have significant leverage at upcoming conferences. Therefore, governments now work extensively within their regional organizations to ensure as much consensus as possible is reached on an issue before an upcoming WRC.

However, it is still rare to see an absolute position supported by all members of a regional organization. This was readily apparent at WRC 2000 during the 3G debate. For example, CEPT came in with a very strong position in favor of identifying spectrum at the 2.5-GHz band for 3G services. However, Russia and several other Eastern European countries that are CEPT members loudly argued against the identification of any additional spectrum for these services at the conference. Hence, a clear rift existed between all of the CEPT member states that impacted the ability of CEPT to achieve its goal of having the 2.5-GHz band identified as the sole frequency band for use for 3G services.

Similarly, the United States did not support the CITEL position to identify spectrum at 1.8 GHz for 3G services, instead proposing a multiband discretionary approach to the identification. Before and during the conference, however, the United States actively worked to gain the support of the CITEL, which was thought necessary to convince the rest of the world of the merit of its proposal. Over the course of the conference, the United States was successful on almost all accounts. However, because the United States did not have the support of its region in advance of the conference, it had to expend valuable resources convincing these countries of its position during the conference.

Despite some dissension within regional organizations, their participation in the radiocommunications spectrum allocation process will continue to be integral. This is in large part because of the trading blocks that have developed between countries within the same region (most notably in this situation for telecommunications equipment) and the need for closer spectrum coordination between neighboring countries.

The international spectrum allocation process

As discussed, the role of the WRCs is to allocate radiocommunications spectrum to individual services on both a global and regional basis and to adopt corresponding technical rules that ensure the successful operation of radiocommunications systems in these services. These conferences may address any service throughout the entire radiocommunications spectrum, dependent on the agendas set by the ITU Council, based on the WRC recommendation. In addition, there are also regional radiocommunications conferences, which meet as necessary and have a restricted agenda devoted to specific services for the ITU region concerned [12]. Based on the agreements reached at these conferences, the ITU publishes the international Radio Regulations, which include allocations and technical rules for radio operation for each of the three regions of the world.

The ITU's WRC designates spectrum allocations on a primary or secondary basis. These types of allocations can be defined as follows:

- A primary allocation grants a specific service or services priority in using the allocated spectrum. When there are multiple primary services within a frequency band, all have equal rights to operate free from harmful interference. A station, however, has the right to be protected from any others that start operation at a later date.
- A secondary allocation is a grant made for services that must protect all primary allocations in the same band. Services operating in

secondary allocations must not cause harmful interference to, and must accept interference from, primary service stations. All secondary service stations have equal rights among themselves in the same frequency band.

At the WRCs, countries must agree on set services in which to allocate frequency bands and any necessary technical rules. In general, radiocommunications spectrum will only be allocated for a service if it is on the agenda for the relevant conference. The agenda sets the framework for the issues that will be addressed at each conference. Although they are tentatively discussed years in advance, agendas are finalized by the ITU Council at meetings held around the previous WRC [13]. The reason for such a long lead time is to ensure that there is sufficient time to study the technical issues that are planned for discussion.

Accordingly, it is important to make sure that upcoming issues are on the agenda. Failure to do so could raise the issue of whether a conference is competent to even discuss a specific allocation, especially if the technical work necessary for the conference to make a decision has not occurred through the study group process. Exceptions to issues being addressed that were not clearly set forth on the conference agenda are rare. The most notable recent exception was the NGSO FSS issue. The WRC 1992 agenda did not have an unambiguous agenda item for inclusion of the issue of the designation of spectrum for NGSO FSS stations on the agenda. Nevertheless, because the United States was able to garner sufficient support for consideration of this item, it was raised, over the very strong objections of several administrations, such as France.

However, there has been a recent movement, primarily by Europe, to try and make WRC agendas more flexible. For example, at WRC 2000, a tremendous amount of effort was placed by the Europeans in making sure that there was an agenda item for advanced mobile communications services. Because such services are not defined by the ITU Radio Regulations, a reference to this category provides a broad mandate for future conferences in terms of their jurisdiction [14].

Spectrum may be allocated to service globally or in one or more of three geographic regions created by the ITU for this purpose. These regions are:

- Region 1, which includes Europe, the Middle East, and Africa;
- Region 2, which includes the Americas;

- Region 3, which includes the Asia-Pacific region.

As discussed, rarely does a vote on whether to create an allocation or adopt a technical regulation occur at the WRC. In general, it is a body that works on the principle of agreement by consensus. Accordingly, most allocations are determined based on compromises among the member states. This may result in substantial horse trading, where one member state agrees to support another on an issue that is of little importance to them in exchange for support on an item of greater importance. In addition, other compromises may be reached, which may or may not involve telecommunications issues and which may in fact be politically based. Accordingly, a key to success at WRC is to understand the position of the other ITU member states before the conference, determine what key issues (both in telecommunications and outside of telecommunications) are for those member states, and begin discussions on possible compromises before and during the WRC process. For example, a government may trade its support on a WRC issue for resolution of a broader trade issue. Such negotiations usually occur up until and including the last day of the conference.

Technical issues

WRCs also set the accompanying technical rules for use of the allocation. These take the form of footnotes, recommendations, and resolutions. accompanying technical rules are not consistent with the planned uses of the spectrum, then the resulting allocation may be close to meaningless for the proposed use. Accordingly, the accompanying technical rules are as important to many interests as the spectrum allocations themselves. In addition, technical rules may be sought for adoption to impede the use of a specified frequency band for a specific use. This may be based on purely technical rationale or may have underpinnings of anticompetitive behavior.

Today there are more than 40 radio services recognized by the ITU Radio Regulations, such as the MS and FS. Over all, allocating additional spectrum to an existing service is a difficult process in light of the congested state of most spectrum. Because most of the radiocommunications spectrum is heavily used, it is often difficult to obtain changes in existing allocations unless alternative spectrum exists for these services and such a move would involve little or no expense to the existing users.

However, it is even more difficult to obtain spectrum for a new type of service. There are several reasons for this. First, the creation of a new radio service means that there are more services competing for a limited amount of radio spectrum. Second, competitors of the proposed new service may try, for anticompetitive reasons, to stop any proposed allocation to retard the deployment of the service. In addition, in most cases new uses of the spectrum will fall into one of the baskets of services (e.g., FS, MS, or AMSS) that have already been created.

Due to both the difficulty of allocating spectrum to a new service and the broad definition that services generally have, the use of the term *identification* of spectrum has come into favor. This is the terminology that has been used for the spectrum marked for use for 3G services at WRC 1992 and WRC 2000. Essentially, this term denotes a preference for the use of certain frequency bands for a set use. However, the identification of spectrum has no regulatory status under the ITU regulations and is not even a defined service. Accordingly, there is no guarantee of protection against harmful interference by other services operating in the band in accordance with the ITU Radio Regulations.

Advocates of the use of the identification of spectrum believe that it provides global guidance on the appropriate use of a specific frequency band when global communications services are at issue. In addition, such an approach is often seen as a way for countries to dictate the use of set technologies or usage of frequency bands.

However, the use of identifying spectrum has not always been successful. For example, at WRC 1992, a significant amount of spectrum allocated to the mobile service was identified for use by 3G services. Despite this action, the United States made a decision to use this spectrum domestically for PCS, a form of second generation cellular services. In letters to the U.S. government, the EU contended that such use violated the ITU Table of Frequency Allocations, but the United States has consistently dismissed this claim as baseless. Because the ITU Table of Frequency Allocations and the Radio Regulations provide an identification of spectrum with no regulatory status, it is clear that the U.S. position is legally defensible.

Another option to ensure the operation of a new technology or system is to seek the designation of spectrum for a particular use at a WRC. Like the identification of spectrum, this is a nonbinding principle under ITU Regulation, as the term *designation* is undefined by the ITU Regulations. What a designation does do, however, is provide for use by a specific technology or type of system by adopting technical rules for use of the relevant frequency band, which ensures its successful use. Such was the case of the NGSO FSS designation at WRC 1997.

An overview of the WRC process

The WRC process is fairly complex, but at the same time generally predictable. Within the confines of the WRC preparatory process, the preparations of the regional organizations who participate in the WRC process, and the WRC itself, there are certain set. These procedures and meetings, discussed next, are supported by the administrative staff of the ITU-R.

Traditionally, WRCs were only attended by member states (and in rare instances, private-sector representatives on member state delegations). However, this has changed in recent years. Today, more and more private-sector members of the ITU can participate in their own right, as well as on member state delegations, to the WRC preparatory process meetings and the WRC process itself (as well as, depending on the regional organization, to the regional organization) based on the rules of the member states from which they are affiliated. However, because the WRC is a treaty-making body, members may not vote on any matters in that body. Nevertheless, the presence of private industry at all ITU meetings has dramatically changed the dynamics of the meetings, whereby the participants now outwardly address commercial issues. It is important to recognize that private-sector participation in the ITU costs money. With the recent economic downturn, many companies have cancelled their memberships. This means that more and more companies will have their prime form of participation occur through serving on country delegations and not in their own right.

As discussed, the agenda for each WRC begins to be set several years in advance. The agenda for the conference is only finalized at the time of the prior WRC and must be approved by the ITU Council [15]. At each WRC, a separate committee that considers items for future conferences is formed. Often WRCs can be quite contentious because of the importance of the issues being addressed. For example, quite often the issues on the table will impact hundreds of billions of dollars of business. Because of its impact, commercial and government interests often expend vast resources at the WRC and in the preparatory process to ensure that they are successful at a WRC.

In addition, the meetings tend to be time pressed because there are generally more items being advocated than a single or even multiple conferences are capable of addressing in the 4 weeks that are allocated to them. Accordingly, quite a bit of compromise occurs on what items should be included in the next conference agenda.

By being identified for discussion at an upcoming WRC, an issue is likely to begin to be addressed by the relevant ITU-R study group during the study group process. Study groups may be broken down into smaller subgroups to address narrow issues. In these meetings, technical and other experts of interested member states will meet to examine the relevant agenda item and issue relevant technical reports for consideration by the upcoming conference. These groups will likely meet several times during the relevant study group period leading up to the relevant work and may also meet via correspondence on a more frequent basis. Each member state has its own rules on how ITU members can participate in these study groups. In many countries, such as in the EU and Canada, members participate fairly freely in coordination with their member states.

During the period leading up to the WRCs, individual countries will try to determine their own domestic positions on the upcoming conference. In addition, these member states will often meet within the confines of their own regional groups, such as CITEL and CEPT, to determine regional positions for the conference. Other negotiations may also occur among countries that are trying to set the groundwork for the upcoming conference. For example, the United States will arrange a large number of bilateral discussions in the months immediately preceding the conference in an effort to garner support for its positions at the upcoming WRC. Many times, these pre-WRC efforts are as contentious, if not more so, as the WRC itself.

In addition, several meetings occur that most member states will participate in to prepare for the upcoming conference. These meetings are called *conference preparatory meetings* (CPMs). At these meetings, member states and private-sector members try to begin to focus the upcoming conference so that their goals can be achieved, with both actively participating in the proceedings. Specific committees (and accompanying subcommittees) are created to address the relevant items on the upcoming conference agenda. Over all, these meetings produce a report that is taken into consideration by the WRC, although it is not binding. The CPM is considered a good forum from which results can be utilized to gauge the outcome of the WRC.

Similar to the CPM, the WRC is broken into multiple committees and subcommittees that mesh with the conference agenda at the plenary session by the chair of the WRC. These bodies are generally broken into multiple working groups to work out the narrow issues of each agenda item. Participants at the conference may include both member states (and private-sector members who comprise member-state delegations) and members. However, as discussed, members may not vote at a WRC.

In addition to these formal discussions, a wide range of private negotiations between member states occurs throughout the conference. In some cases, these negotiations simply focus on the merits of a particular party's arguments. In a large number of other cases, trading support on particular issues may be involved. In still others, the negotiations may also focus on issues that are not solely within the confines of the conference, such as trade issues that may be of particular interest to one of the parties. Further, in some cases negotiations among internal delegations to the conference may occur because a member state may choose to change its position during the conference. This will be discussed further in Chapter 5.

Each working group attempts to resolve all issues that are placed within its mandate before issuing a report to be considered by the subcommittee to whom the group reports. However, quite often, the most contentious issues will not be agreed to at these meetings. The subcommittee will then make its best efforts to find a resolution to these same issues and confirm that there is agreement on as many issues as it has jurisdiction on before submitting these conclusions to the next level of hierarchy. This continues all the way up through the plenary session. For the most part, a significant number of issues are resolved at the committee level and below. However, the most contentious issues are generally left for the last few days of the conference to be decided at the plenary session. Because of the lack of time of the waning conference, the chair of the conference is generally successful in pushing through what can be considered a consensus position that is acceptable to most of the interested parties.

Conclusion

To date, most experts believe that the WRC process has been largely successful in its allocation responsibilities. There has never been an example of a WRC that has failed in its mission (although issues may be deferred to be addressed at a subsequent conference).

However, many governments and private-sector interests are beginning to question whether the WRCs are able to handle the growing use of the spectrum and the fast-changing technologies. This is a reasonable concern, as agendas are set for upcoming conferences up to 3 years in advance. Accordingly, conferences are generally not addressing the most current technologies. This creates a regulatory lag in addressing cutting-edge issues.

Further, many spectrum participants believe that the WRC process is too political in nature and does not focus sufficiently on the technical merits at issue. This concern has been outstanding throughout the life of the ITU. It is unlikely that the spectrum regulatory process will ever be devoid of political considerations because of the large stakes that are involved. In fact, the process is likely to be even more political as the spectrum resource becomes more congested and the financial stakes continue to increase.

Other critics contend that the ITU process is overly regulatory and stifles technological innovation. These critics believe that a less formal procedure is required with more flexibility worked in. On the other hand, there are advocates of an even more formal regulatory process—one whereby the specific use of the spectrum is also determined by the WRC process. This is best demonstrated by the push for the identification of spectrum for certain uses, such as 3G.

Further, over the next few years, one can anticipate that the WRC process will come under increased scrutiny by member states and by private-sector members as spectrum becomes scarcer. In addition, private-sector members will begin to demand a greater role in the spectrum allocation process, especially as they continue to pick up a growing amount of the financing of the ITU. It is unlikely that this movement will be well received by governments, which will likely see the increasing presence of the private sector in the spectrum-allocation process as a threat to their sovereignty and governmental rights. Despite these criticisms, most experts agree that the ITU plays an integral role in ensuring the rational allocation of spectrum globally. Accordingly, the ITU can be expected to continue to be the lead organization in the global spectrum-allocation process for many years to come, although certain processes and procedures may need to be revisited to ensure its effectiveness.

Endnotes

[1] Dawson, Keith, "FCC/Nextwave: Face Is Saved All Around," October 17, 2001, available at http://www.commweb.com/article/COM20011017S0002.

[2] Mark, Roy, "FCC to Refund Deposits for Nextwave Spectrum," March 27, 2002, available at http://dc.internet.com/news/article.php/2101_999591.

[3] See "U.S. Sees Spectrum Proposal as 'Bridge' at Upcoming Conference," *Communications Daily,* March 20, 2000.

[4] http://www.itu.int/ITU-R/information/brochure/wrc/index.html.

[5] ITU Constitution, Article 44.

[6] See Allison, Audrey, "Meeting the Challenges of Change: The Reform of the International Telecommunication Union," *Federal Communications Law Journal*, Vol. 45, 1992, pp. 491–519.

[7] ITU Constitution, Articles 1 and 12.

[8] See Allison, Audrey, "Meeting the Challenges of Change: The Reform of the International Telecommunication Union," *Federal Communications Law Journal*, Vol. 45, 1992, pp. 491–535; see also http://www.itu.int/ITU-R/information/brochure/rrb/index.html.

[9] http://www.itu.int/ITU-R/information/brochure/br/index.html.

[10] http://www.itu.int/ITU-R/information/brochure/rsg/index.html.

[11] http://www.itu.int/ITU-R/information/bureau/.

[12] http://www.itu.int/ITU-R/information/brochure/rrc/index.html.

[13] See http://www.itu.int/aboutitu/structure/index.html.

[14] http://www.itu.int/md/meetingdoc.asp?type=sitems&lang=e&parent=S02-CL-C-0072.

[15] http://www.itu.int/ITU-R/information/brochure/wrc/index.html.

5

Domestic Regulation of Spectrum, Part I: International Representation

Overview of the domestic regulation of the radiocommunications spectrum resource

This chapter and Chapter 6, which focuses on the domestic regulation of the radiocommunications spectrum, complement the discussion in Chapter 4 on international regulation of the radiocommunications spectrum resource. The concepts contained in these three chapters will provide a basis for many of the discussions on the conflicts that occur over spectrum allocation and usage later in this book.

With regard to domestic regulation of the radiocommunications spectrum resource, domestic regulators are generally responsible for at least the following responsibilities:

- *International representation.* Active participation and representation of their country in the international arena in the role of spectrum regulation and allocation.

- *Domestic spectrum allocation.* Allocation of frequency bands domestically to specific radiocommunications services.

- *Determination of the use of specific frequency bands.* Identification or designation of discrete frequency bands for specific uses.
- *Domestic assignment of spectrum.* Assignment of spectrum to specific users and authorization of the use of the spectrum.
- *Implementation and enforcement of technical and operational rules.* Establishment, implementation, and enforcement of the technical and other rules governing the use of the spectrum resource.
- *Regulation of secondary markets.* The regulation of secondary markets for spectrum authorizations and assignments.[1]

This chapter focuses on the role of the domestic regulator in formulating international positions on spectrum allocation and regulation. Chapter 6 further examines the role of domestic regulators and focuses on the specific areas of responsibility that almost every national regulator has with regard to the domestic regulation of the spectrum resource, including the allocation, assignment, and use of the spectrum resource. In addition, both chapters explore the role that the private sector and other governmental bodies play in each of these areas. These chapters further provide examples of some of the processes that are used by the regulator to fulfill its responsibilities in the area of the regulation of the radiocommunications spectrum resource.

This chapter and Chapter 6, however, do not endeavor to cover each country's individual domestic processes. Instead, they provide an overarching view of the majority of these processes. The advocate that is looking to enter the market must work to understand the individual rules and regulations governing spectrum regulation in the relevant country. This is best accomplished through research and fact-finding missions to the country to meet with the government and other experts and through the hiring of in-country attorneys and consultants.

Participation in the international arena

One of the most important roles that the domestic regulator plays in spectrum regulation is its participation in the international spectrum arena,

1. An increasingly important domestic responsibility is national security and emergency preparedness. The regulator and the defense department of the relevant country often handle this jointly.

such as the WRC and other international and regional conferences where spectrum allocation and use are a focus. An overview of the international process was provided in Chapter 4. In general, international participation by regulators includes actively participating at international conferences by introducing proposals, negotiating solutions, and voting when required. These actions are the result of individual country positions being developed prior to the international meetings. Each country has a different process for developing its positions for these meetings and also has additional government, and, in some cases, private-sector participants. Here is an overview discussion of these processes and the role of the participants in them.

Domestic participants in the international process

There are two broad types of participants in most countries' preparations for and participation in the international radiocommunications spectrum arena. These are government and private-sector participants. What follows is a brief discussion that builds on the participant discussion contained in earlier chapters. It is important to understand the role of these two broad categories of participants, as they are critical to understanding the workings of the domestic international preparatory processes on spectrum-related matters.

Governmental participants

In almost all nations, multiple governmental entities have a role in an individual country's participation in the international process governing spectrum regulation. If nothing else, these entities include the government agency responsible for foreign affairs of the state, such as a Department of State or the Ministry of Foreign Affairs. In general, this agency participates through an office that has some subject matter expertise in telecommunications. For example, in Australia, the Australian Department of Foreign Affairs and Trade is represented at WRCs and similar meetings, as are other branches of the Australian government [1]. Such representation may also occur through or be supplemented by the country's diplomatic corps stationed in the country in which the meeting is occurring. The role of the foreign affairs representative is to ensure that the government's views from an international perspective are included in any international negotiations.

In addition, the actual subject-matter expert agency or agencies, such as the regulatory agency or the Ministry of Communications, also generally participate in the international process. In this regard, usually a branch of the agency responsible for international matters or spectrum issues takes the lead for the agency. For example, in Mexico, different departments of the regulator COFETEL may address different issues [2]. This agency acts as the subject-matter expert for the meeting. Similarly, in Brazil, the regulator ANATEL also has a key role in the international process through active participation and input. In many countries, the views of the foreign-affairs department may trump the views of the subject-matter agency. This is because of the broad-reaching impact one issue may have on the outcome of subsequent issues in other areas, such as the environment and defense.

Other governmental agencies, such as the military, the agency responsible for aviation, or other agencies that are major users of the radiocommunications spectrum resource may participate in their own right or may be represented through another governmental agency. For example, in the United States, the National Telecommunications and Information Administration, a part of the U.S. Department of Commerce, represents all government interests in telecommunications and participates on their behalves in the international process [3].[2] In addition, a limited number of other governmental representatives, including those most interested in spectrum issues, such as the U.S. DOD and NASA, directly participate in the international processes related to the radiocommunications spectrum resource.

There may also be participation in the international arena by governmental agencies that have jurisdiction over promoting industry in the individual country. An example of this is the Department of Trade and Industry in Canada, which actively participates in the international process on behalf of the Canadian government, with an eye towards promoting Canadian industry [4].

Private sector participants: the telecommunications industry

Other key participants from an individual country can include members of the private sector, such as satellite operators and other network operators, telecommunications service providers, and telecommunications equipment manufacturers. These private-sector participants may participate

2. As discussed subsequently, the FCC has responsibility over the regulation of the commercial portions of the radiocommunications spectrum, while NTIA coordinates and regulates the government-controlled and government-used spectrum. Both agencies regularly coordinate with one another on these issues. When an issue has international ramifications, the Department of State is also consulted.

either directly or through trade associations, such as the Wireless Communications Association International, the Telecommunications Industry Association, or the Cellular Telecommunications Industry Association.

Also, unlike the governmental participants, the private-sector participants may participate in the international process through multiple governmental processes and on multiple member state delegations to international conferences. Therefore, a single company with operations in multiple countries, such as Motorola, may participate in processes in many countries, including the United States, France, and Brazil.

In some cases, as previously discussed, private-sector participants may participate in their own right in these processes as a member of an organization, such as in the ITU. Many countries, including the United States and Canada, allow such direct participation. Many companies have taken advantage of this ability, including AT&T, Nortel, and Qualcomm.

Members of the private sector who are interested in spectrumissues being raised at the international level often expend substantial resources participating in individual country preparatory processes. Because of the specialized nature of spectrum regulation, companies whose business is heavily dependent on the outcome of such issues, such as Hughes' satellite division and telecommunications equipment manufacturers such as Nortel, often have large departments devoted to participating in spectrum-related matters and also hire consultants to assist them. Smaller companies or those companies with less direct involvement may only have one or two employees assigned to such work or may rely entirely on consultants or trade associations to represent their interests.

Consumers

Rarely are consumers direct participants in the international process. This is because of the cost involved in active participation and the often-mistaken belief on their part that their interests will be represented by existing participants, such as their service provider or the government. Because of the lower cost and familiarity with the process, consumers are more likely to play a role in the individual domestic processes that lead up to international meetings and determine the individual country position for those conferences. This participation is generally achieved through commenting in public proceedings or direct lobbying. In many countries, consumers are also active in the preparatory process through business councils and other trade associations. This type of mass participation is often dramatically less resource intensive than the direct participation of one company in the process and may be sufficient to safeguard the consumer's interests—which generally are focused on issues of cost and availability of an existing or new service.

The domestic preparatory process for international meetings

Almost all countries have an established procedure for developing domestic positions for international conferences and negotiations. In general, in competitive telecommunications markets, such as Australia, Brazil, Japan, and France, the private sector can directly participate in the preparatory process for major spectrum-related meetings [5]. However, in markets with very limited or no competition (including generally non-WTO member countries), such as Cuba, the private sector can play no official role in the preparatory process. In such markets, the government generally makes and implements the positions that will be taken at the upcoming international meetings, without any or with only minimal private-sector input. In such markets, private-sector participants are generally reduced to providing inputs into the process through the lobbying process.

In the more open markets, a multiyear process is often utilized by the government in preparing for each WRC, as well as in the preceding international and regional conferences and bilateral negotiations. Significant resources are often expended in the preparation process by both the government and the private sector. As discussed, such resources may include costs, employees, advertising, travel funds, and the development of technical studies. These costs are rationalized by the importance of the impact to the particular participant. For example, if Teledesic was unsuccessful in its efforts to have Radio Regulation 2613 eliminated at WRC 1995 in the 28-GHz band, then it would not have been able to operate its proposed system because of the potential for interference between its operation and that of the GSOs also in the same frequency band. Accordingly, Teledesic was able to justify expenditures of a large amount of resources (both capital and others) in its endeavors, including its participation in the preparatory process in multiple countries.

The domestic processes used in WRC participation, as well as in the meetings leading up to the WRC both at the ITU and other international bodies, may involve the initiation of rulemaking proceedings, combined industry and government meetings, and direct lobbying of staff-level and senior governmental officials by the private sector [6]. In addition, in many cases, the government will prepare its own internal positions for the upcoming meetings, sharing only certain pieces with private industry. Even in the most open of markets, such as the United Kingdom, this is the case because governments may have national security interests at stake which they cannot share with private industry. In addition, governments may trade a position on a radiocommunications spectrum issue in order to obtain a concession in another area, such as energy, but not inform industry about the planned concession. Further, the government often excludes industry participation in government-to-government bilateral agreements on similar grounds. When this occurs, industry becomes increasingly reliant on having the government as advocate of its position. Accordingly, it is quite common for companies to extensively lobby the government agencies participating in the bilateral agreement on issues of concern to them that they want to be discussed during the meetings.

One of the most established and complex of these domestic preparatory processes is that of the United States. The United States starts its preparatory process for each WRC almost immediately after the last WRC has concluded. The United States' process is bifurcated, with separate processes being used to develop private-sector and governmental positions for the upcoming conference. More specifically, the FCC runs a multiyear committee meeting process with the private sector, which results in a report that outlines industry's views on the upcoming WRC agenda items [7]. At these meetings, the private sector and other participants, including governmental agencies, will table proposed position papers and technical studies for discussion. Over the course of the negotiations in these meetings, consensus positions are reached and included in a final report, along with an identification of those positions on which a consensus could not be reached. In general, the FCC endeavors to obtain consensus on all issues, even if that means having separate meetings with interested parties outside of the IAC process.

Simultaneously, NTIA chairs a governmental meeting process that results in a governmental position on the WRC agenda items [3]. In the NTIA process, governmental users of the spectrum, including agencies such as the FAA, the U.S. DOD, NASA, and the U.S. Coast Guard meet regularly to come up with proposed positions for the upcoming WRC. Government agencies often have conflicting views for the use of a particular frequency band. These conflicts are quite often politically resolved.

The reports of the FCC and the NTIA processes are often inconsistent on the more contentious issues, especially when the reallocation or redesignation of government spectrum or a proposal raising the potential of harmful interference into an existing or planned government service is involved. However, during both meeting processes, members of these groups will have extensive discussions to try to resolve as many issues as possible and have as consistent an approach as their own interests will allow.

Each of these reports is tabled with the U.S. Department of State's *Office of Communications and Information Policy* (CIP). Upon receipt of the reports, CIP works to combine the positions of both reports and works to find consensus positions on those areas where there is disagreement (although quite often, where it is apparent that areas of disagreement exist, behind the scenes discussions will already be taking place) [8]. At times, however, the only way for CIP to resolve an issue is to take a unilateral decision. However, in many more instances, CIP will work to find consensus among the parties. The result of the two processes is the creation of the U.S. position to the WRC. The United States, however, does not finalize its positions or draft its final proposals for the conference until it forms its conference delegation, as such discussions are limited to actual delegations members (with the exception of the government). This is because delegation members, as opposed to the general public and industry at large, are subject to strict confidentiality restrictions on sharing such information. It is important for the delegation to know that information will not be leaked to other negotiators at the conference. Accordingly, confidentiality of communications is critical.

A good example of an effort to find consensus was in the preparatory process for WRC 2000 concerning which frequency band should be the one that the United States recommends to be identified for use by 3G services. Although both the government and private-sector preparatory processes agreed that the 2.8-GHz band, which was used by the FAA, should not be identified, there was a split over whether the 1.8-GHz or the 2.5-GHz band should be identified. As discussed previously, the 1.8-GHz band was a band used by the U.S. military, and the 2.5-GHz band was used by the MMDS community. Many interests, in fact, believed that the United States should not take any position. However, the 3G users argued, and many others believed, that the United States would lose out in the discussions on the future of mobile telephony if it did not take some position. Accordingly, the U.S. Department of State chaired an informal group with 15 private-sector and government representatives to resolve this deadlock. After many straight weeks of contentious meetings and side negotiations, the United States was able to come up with a consensus position on this issue; it would propose to the conference that both the 1.8-GHz and the 2.5-GHz bands be identified for use for 3G services, with each country having the right to choose which band was the right one for it to offer such services in [9]. Ultimately, this was the position that was adopted at WRC 2000. Many interests believe that this position carried the conference because the United States had to hold its own domestic mini-WRC in advance of the conference to reach this consensus.[3]

Similar preparatory processes are often used by national governments for other relevant meetings, such as those held by CITEL and APT. More often than not, however, those meetings are prepared for domestically in a less formal method, with just a couple of preparatory meetings held on the domestic level and, at times, fewer interested participants. However, as the WRC nears, regional meetings increase in importance and may be subject to more complex preparatory processes by individual governments.

Of course, some governments may have a less intense focus on the preparatory process than the United States does for international spectrum issues. In smaller countries, such as Malta, the process may only focus on a few issues of concern to the country and its industry because they do not plan to actively participate in discussions on issues not directly concerning the country. In such cases, much less formal processes may be utilized, and the government and the private sector may play a less active role in regional and international meetings (except on issues of immediate concern). Other countries in a similar position may still want to prepare on discussions of issues that are not of direct relevance to them. In this manner, they may be able to trade support on such an issue, to gain support on an issue of concern at the meeting, or for an unrelated goal.

The delegation to and attendance at international conferences

Governments participate at international radiocommunications conferences in the form of national delegations. These delegations are important because it is through them that a country determines its ultimate position during a meeting on a particular issue. The national delegations speak on behalf of the country and take votes.

3. The United States process is somewhat unique in that it appoints an ambassador to head the U.S. WRC delegation. That head of delegation is only appointed several months before the start of the WRC. Often, this delegation head may have little or no experience in telecommunications and has only a very short period to obtain the subject-matter knowledge and credibility necessary to effectively lead the delegation.

Accordingly, a key part of the domestic regulation of the radiocommunications spectrum resource is the rules and regulations governing the ability of the private sector to participate at international conferences. Although governments generally have finalized positions prior to an international conference, governments are also likely to change these positions during the course of an international negotiation in order to obtain a concession in another area of interest. Accordingly, the private sector has increasingly begun to recognize the importance of participating in the formal delegation to international conferences in order to attempt to influence the government on their position up until the last possible moment.

Further, serving on the country delegation provides the participant direct access to its own government, in an attempt to ensure a favorable outcome on its own issue(s) of concern. In addition, being on an official country delegation often is seen as a sign of status and generally provides greater access to foreign delegates, who may be more circumspect of holding discussions with a purely private-sector participant. As these benefits have begun to be recognized, private-sector participation on national delegations to international radiocommunications conferences has increased exponentially.

It is interesting to note that up until the past few years, private-sector participation on national delegations was limited to a small number of members. However, that has changed dramatically, and today most competitive telecommunications markets allow at least some private-sector participation. The impetus for this change was the introduction of competition into the world's individual telecommunications markets. Previously, almost every telecommunications market had primarily government-owned and government-operated telecommunications service providers. As the telecommunications markets around the world have opened to competition, governments faced increasing pressure to also open the international meetings that directly impact the market to private-sector participation.

In some countries, the number of delegation members is still extremely limited, so that not all interested parties can serve. In these cases, delegation members who are believed to represent a wide variety of interests are often selected . This is in direct contrast with countries such as France and the United States, where almost unlimited participation is allowed and the process can appear unwieldy. There is a benefit to such an approach. This is that the delegation is large enough to cover all of the relevant meetings on issues of concern and to hold meetings with as many delegates from other countries as possible in an effort to obtain support for key issues. This ability to engage in direct discussions with many other delegations is extremely important because each country has one vote in the international process—so each vote is critical.

One interesting note to remember is that although private-sector participants may serve on a delegation, they are not allowed by most governments to vote directly or speak on an issue on the floor without government sanction. In this regard, the private-sector member is an advisor to the government decision maker and rarely, if ever, the actual decision maker. Accordingly, industry may find itself in the untenable position of counting on government support throughout a meeting, only to be traded away at the last minute in favor of an issue of greater concern to the government.

Another possible method for private-sector participation at international meetings on radiocommunications issues is to participate as a private-sector member. This is possible in many organizations, including the ITU and CITEL [10]. Such an approach may allow the private-sector participant the ability to submit papers to the meeting and also speak from the floor. However, the ability to vote is still not allowed. In addition, some countries may have restrictions on the ability of their private sector to play such a role or restrictions on the scope of activities that the private sector can engage in while participating as a private-sector member. For example, some governments may not allow a private-sector participant, even when participating in their own right, to introduce any papers that are inconsistent with the country's position on the same issue. Such restrictions exist to protect the interests of the individual nation, albeit at the expense of the private-sector member.

It is clear that the role that the private sector is permitted to play at international radiocommunications meetings is continuing to grow. However, it is clearly unlikely that the private sector will ever have the voting rights that member states are afforded at the ITU.

Endnotes

[1] See http://www.aca.gov.au.

[2] See http://www.cft.gov.mx.

[3] http://www.ntia.doc.gov.

[4] http://www.ic.gc.ca/cmb/welcomeic.nsf/558d63659099294285256488005215 5b/7cf90ec81c6762a105256b03004eec34!OpenDocument (discussing Industry Canada's role in leading the 2000 Canadian WRC delegation).

[5] See, for example, http://www.aca.gov.au (discussing Australia's International Radiocommunications Advisory Committee, which is charged with developing Australia's positions on WRC issues).

[6] See, for example, http://www.fcc.gov/ib/wrc-03/pnotices.html (outlining public notices issued by the FCC on WRC 2003).

[7] http://www.fcc.gov/ib/wrc-03/.

[8] See, for example, http://www.state.gov/e/eb/cip/c293.htm.

[9] See "U.S. Sees Spectrum Proposal as 'Bridge' at Upcoming Conference," *Communications Daily,* March 20, 2000.

[10] See http://www.itu.int/cgi-bin/htsh/mm/scripts/mm.list?_search=SEC.

6

Domestic Regulation of Spectrum, Part II: Allocation, Assignment, and Use

Overview

This chapter focuses on the domestic regulatory regime governing the spectrum resource and is a supplement to Chapter 5. In this chapter, the discussion centers largely on issues concerning the domestic allocation, assignment, and use of the radiocommunications spectrum resource. In addition, this chapter also focuses on the implementation and enforcement of technical and operational rules, as well as touching on the subject of the regulation of secondary markets for spectrum authorizations and assignments, which will be discussed at much greater length later in this book.

In order to fully understand the domestic regulation of the radiocommunications spectrum, it is important to understand who has the responsibility for awarding access to the spectrum resource and who has the responsibility for regulating the use of the spectrum in each country. In many countries, a separate administrative agency or office has responsibility for the regulation of the spectrum resource (with regard to allocation and authorization issues), but the regulator handles policy and similar matters. For example, in the United Kingdom, three different agencies were involved with such regulation. These included the *Department of Trade and Industry* (DTI), the agency with responsibility over policy issues; OFTEL, the independent regulator with responsibility for telecommunications regulation; and the Radiocommunications Agency, an arm of the DTI charged with frequency management, control, and assignments and authorizations [1]. Such an approach to regulation creates the need for extensive coordination among the interested governmental agencies to ensure that they are synchronized with regard to policies and actions.

In other countries, all regulation is handled directly by a single reglator, such as Brazil's National Telecommunications Administration (ANATEL) [2]. This regulator is responsible for all aspects of the domestic allocation of spectrum—determining the appropriate use of the spectrum, assigning spectrum, implementing and enforcing technical and operational rules for radiocommunications uses, and developing and implementing overall public policy. A major benefit of the single regulator approach is that only internal coordination on telecommunications matters is necessary before taking action and the regulator is also aware of the broad implications to the telecommunications market of its actions.

One drawback, however, to not having a separate agency to handle radiocommunications issues is that these issues may not be of particular interest to the overall regulator in comparison with other issues that it is addressing, such as interconnection. This may delay the needed licensing or other regulatory actions that are required in the spectrum arena while the regulator handles issues of greater concern. In addition, certain subject-matter expertise may be less available when there is no single entity charged specifically with regulating the spectrum resource.

In either case, however, rarely does a regulator work in a vacuum. Regulators will often be influenced by other branches of governments in their decisions, even in a governmental structure that attempts to preserve the independence of the regulators. This influence often comes most directly from the head of the state that appointed the regulators or from the government body (often the legislature) that controls the agency's purse strings through the budget process. For example, the legislature may send letters to the regulator clearly stating their views on issues pending at the agency or call public hearings on these same issues in order to further demonstrate the importance of their view on the relevant issue to the regulator. In theory, regulators should be immune from such actions and pressure. However, in reality, such efforts may end up being an important factor in any regulatory decision. Unfortunately, it is often hard for an aggrieved party to demonstrate that a particular decision of the regulator is politically or otherwise improperly motivated. For example, in many markets regulators still have close ties to the former monopoly service provider. In such cases, a decision may be made in favor of that provider based strictly on such preexisting relations to the detriment of other parties. However, it may be very difficult for the aggrieved party to demonstrate that this is the reason for the decision. Accordingly, in such situations, the aggrieved party may be left with no form of recourse.

Further, whatever the regulatory structure, another separate entity may be responsible for the allocation and assignment of spectrum to government agencies. For example, in the United States, the NTIA was established to handle the assignment of spectrum to government users [3]. Accordingly, in such structures, close coordination between the regulators responsible for government and private-sector use is necessary. Often, debates on whether the private sector or the government should utilize spectrum may occur between the agencies. In some cases, a shared approach may be adopted, while in others, one side may win out over the other.

In either event, in most countries, the allocation and assignment of spectrum to government uses is not subject to the transparency and other administrative processes that govern the regulatory regime for private-sector use of the spectrum resource. In fact, even the WTO Agreement, which is the main agreement on trade in telecommunications services, excludes such actions from the terms of its regulatory reference paper. Accordingly, the private sector often has little visibility and direct influence on the resolution of issues that may directly impact their interests within these forums.

The distinctions in the different bodies that play a role in the domestic regulation of the spectrum resource are important to understand as this chapter addresses the relevant processes associated with spectrum regulation. Each form of regulation has its pros and cons and may directly impact how the processes flow, with added delay often seen in countries where there are multiple agencies with responsibility for spectrum allocation, assignment, and use issues.

Important cornerstones of domestic regulation

In regulating the spectrum resource, domestic regulators are governed by the basic principles of regulation that govern all of telecommunications. In most countries with largely competitive telecommunications markets, this means that regulatory actions are generally made based on a form of public-interest standard that is contained within the statutory mandate for that country. For example, in Brazil, ANATEL was formed for the purpose of ensuring that "services are offered in a fair competitive environment ensuring the maximum benefit for the Brazilian society" [4].

In this regard, ANATEL has the mandate to [4]:

- Foster full and fair competition among service providers;
- Establish conditions to avoid private monopolies;
- Prevent transgressions against the private order.

Due to its subjective nature, determining what constitutes the public interest is often difficult, with many factors considered and weighed against each other. The public-interest determination is often made in conjunction with other required criteria, as established by enabling legislation such as national security. In addition, governments are also bound by principles of administrative law in making the public-interest determination, especially in competitive markets. Accordingly, decisions based on the public interest have to be rational, or they face being overturned on appeal.

Impact of the WTO Agreement

An increasing presence in the regulatory regime of many countries is the regulatory principles encapsulated within the WTO Agreement. To date, more than 80 countries are signatories to the WTO Agreement, with each bound to comply with their country's commitments. This includes almost all markets that are either opening or are open to competition, such as the member states of the European Union, the United States, Japan, Brazil, Ecuador, and the Philippines. While many potentially significant markets, such as North Korea and Saudi Arabia, are not yet members, more and more countries are seeking membership. A good example of this is China's recent accession to the WTO Agreement in 2001.

The WTO Agreement covers all telecommunications services, including spectrum-based services, in a technology-neutral manner. More specifically, the WTO Agreement covers several major areas. The first is the general principles of the entire WTO accords—national treatment for signatories and the availability of an established dispute-settlement mechanism. The second area covered is the individual countries' commitments on the terms and conditions for entry into the telecommunications market. Because the WTO Agreement is an asymmetrical agreement (so that each country has its own set of commitments), it is important to examine the individual commitments of each country carefully to determine what the country is required to implement within its domestic regulatory regime. The third main area is the principles contained in the Regulatory Reference Paper that most WTO Agreement signatories have committed to implement. These include the following [5]:

- The establishment of an independent regulator;
- The creation of a transparent regulatory process;
- Implementation of anticompetitive safeguards;
- Provision for nondiscriminatory interconnection;
- Transparent and fair universal service obligations;
- Nondiscriminatory allocation of scarce resources.

Accordingly, in WTO markets, the governmental process is required to be very visible and predictable. However, differences on how individual governments implement these principles will vary greatly (and, in some cases, not all of these principles were made part of the commitment of the WTO Agreement signatory). Accordingly, it is imperative that the market entrant fully understands the individual processes that exist governing each of these areas.

However, in non-WTO markets, no such guarantees exist. A brief survey of these markets reveals that most, if not all, of these markets do not have a domestic regulatory structure that would be consistent with the WTO Agreement regulatory principles.[1] This is important for an applicant for radiocommunications spectrum to understand because it may make the allocation and assignment process more complicated, as the ground rules for participating will not be clear or necessarily provide for participation on a level playing field. As the next sections on allocation, assignment, and use of the radiocommunications spectrum develop, the principles of the WTO Agreement should be referred back to, especially with regard to countries that have acceded to the agreement.

1. For instance, in Vietnam, the government processes are not transparent in the awarding of regulatory concessions. Similar conditions exist in many other non-WTO markets, including Saudi Arabia and the United Arab Emirates.

The domestic allocation of frequency bands to individual services

As discussed, WRCs are responsible for the international allocation of individual frequency bands to specific radiocommunications services on both a global and regional basis. On the basis of such allocations, individual countries are authorized to make domestic allocations to radiocommunications services. Such allocations do not have to be in conformance with the international allocations, but they are required (with certain limited exceptions) to ensure that operations in these frequency bands do not cause harmful interference into the uses of other countries operating in accordance with the International Table of Frequency Allocations.

In general, countries that have competitive telecommunications markets utilize a public proceeding or process to initiate the allocation of spectrum to a specified use. For example, in South Africa, the government issues public notices on these issues and requests public comment and input. In most cases, this involves the publication of a written proposal by the government announcing its plans to allocate spectrum to a specific use and the reasons for such an allocation. Generally, accompanying technical rules will also be proposed for the use of the spectrum. Interested members of the public are then invited to comment generally in written, and sometimes in verbal, form to the proposal. Often, heavy lobbying on the part of interest parties, each advocating their particular viewpoint, will accompany these proceedings.

The government is next required to review the formal responses to its proceeding and make a determination as to whether to proceed with the allocation of the frequency band to the proposed service on the basis of these comments and on the relevant public-interest standard with which it is required to comply. An appeals process generally is available for parties that may be aggrieved by the resulting decision of the regulator.

Another mechanism that is available to the private sector in many countries is a request by an individual or group of individuals for the government to allocate a frequency band to a specific service. In some countries, the government is required to act on this request (either through issuing a public proceeding addressing the issue or denying the request), while in others, such as the United States, no action is ever required, unless the government wants to act on the request. In general, a government that does not support such an allocation will find a way in which to avoid making the allocation, even if the allocation is consistent with the government's international obligations. At the end of the day, a government's sovereignty will always allow it to make its own domestic determinations on the allocation of radiocommunications spectrum to a particular service within its own national boundaries (although subject to the ITU-treaty mandate not to cause harmful interference to the stations of other countries operating in accordance with ITU allocations and accompanying technical rules).

In an effort to ensure success in the spectrum-allocation process, many advocates will expend substantial resources on lobbying and studies to convince the regulator and other relevant governmental entities of the merit and necessity of the proposal being considered and ultimately granted. In general, lobbying efforts will occur at both the executive and legislative levels, in addition to at the regulator. Furthermore, the advocate may look for additional support from similarly situated companies, consumers, equipment manufacturers, and the like. In this manner, the advocate can try to obtain the political support it may need to assist it in convincing the regulator to take its proposed action. Additionally, political support may be more forthcoming if public-interest groups are enlisted to support the proposed action.

Because the categories of radiocommunications services are so broad, changes to the domestic Table of Frequency Allocations are not a common occurrence. In fact, many new uses of the spectrum resource try to fall within an existing allocation in order to avoid this step. For example, when a new use of the spectrum, called *high-altitude platform service* (HAPS) (a series of blimps circling the globe providing telecommunications relay services), was proposed, this use made an argument that many governments supported in order to avoid obtaining an allocation to a new service. Specifically, the HAPS applicants advocated that HAPS fell within the definition of the FS. The frequency bands that it wished to operate within were allocated to the FS. Accordingly, the only changes that were necessary to obtain authorization for the operation of this service were changes to the technical rules in the frequency bands allocated to the FS where it sought to operate (and to authorize its use to provide subsequent 3G services) [6].

In countries where telecommunications markets are not competitive or are fairly closed, the government process for the allocation of spectrum may be very hard to discern. For example, in some cases, such as Iraq, the government may not officially provide for any public input into such a process. In others, the government may provide for some public participation, but may not be required to consider this comment in its deliberations. Further, unlike in more competitive markets, where there is generally an appeals process available, such procedures may not exist—or, if they do, they may be for all essential purposes a fiction.

The identification or designation of radiocommunications spectrum to specified uses

Another key regulatory function is the identification or designation of spectrum for a particular use. In many countries, this process may be self initiated by the regulator or by a request from the private sector. In either case, a rulemaking procedure, such as discussed earlier, is generally utilized in more open markets with a direct public participation in a transparent process.

Such rules changes generally take two forms when initiated or sought by the private sector. The first is an effort on the part of the private sector to have rule changes enacted to ensure that the proposed use of the spectrum operates on an interference-free basis. This generally requires changes to existing technical regulations to ensure the spectrum is able to support the proposed use.

In other cases, the advocate may seek to have existing users with whom it might interfere removed from the frequency band so that it can enjoy interference-free operation. In addition, the government may also self initiate such changes in order to encourage the use of a frequency band for a specific use or uses.

Changes to rules

As discussed, most new uses of the radiocommunications spectrum fit within an existing radiocommunications service. However, the technical rules authorized in the relevant frequency band may not permit the use or the optimal use of the band by a proposed use. Accordingly, a proceeding may need to be initiated to allow such an operation through either the adoption of new rules or the modification of existing rules. In most countries, such as Germany or Japan, this will occur through a public proceeding, where interested parties can directly comment on any proposed changes to the existing technical rules that will allow a specified use. These proceedings may also examine related issues, such as whether authorizing the proposed use will impact other authorized service providers, and any other impacts there may be.

It is rare that such an effort would not meet with objections from existing users in the spectrum or competitors to the use. Accordingly, most of these proceedings are contentious and often take a long time to resolve.

Further, because of the political nature of these decisions, they are even more likely to be delayed as compromises are sought. Unfortunately, the delay that is inherent in such proceedings often holds up the introduction of new and innovative uses of the spectrum to the public, albeit while providing procedural protections to existing users.

Flexible use

One issue that has arisen in recent years is how specific a government should be with regard to identifying uses in a frequency band. The idea that more flexibility should be afforded in designating spectrum for a use is a growing trend. The idea is to avoid the delay inherent in holding individual proceedings each time a new use is proposed or a technical rule needs to be changed in order to allow the deployment of a new technology. In response to the recognition for the need to allow new technologies to be deployed quickly, several governments have begun to create a more flexible-use approach to the designation or identification of spectrum. This means that the frequency band is not designated or identified for one form of use but can be used by any use that meets the broad technical and operational rules governing that frequency band (and ensuring that unacceptable interference does not result into operating uses). Many advocates of this approach believe that this will allow the market, as opposed to the government, to decide what is the best and most efficient use of the spectrum.

A good example of this was the FCC's decision to allow the use of the 2.5-GHz band for mobile uses, in addition to the FS use by MMDS and ITFS [7]. Prior to this rule change, certain technical regulations governing antennas that were permitted to operate in this frequency band deemed mobile use impossible, although the frequency band had a coprimary allocation for the FS and the MS. After public comment on this issue, the FCC determined that it was in the public interest to revise its technical regulations to allow service providers in the band the flexibility to operate the most appropriate technology, whether fixed or mobile, in order to serve their customers. Other countries, such as those in the European Union and Australia, are also looking to this approach.

Another form of this approach provides broad flexibility to existing operators to offer *ancillary* uses. Ancillary uses can be defined as those uses that are allowed for as a supplement to the existing use. A good example of this is the recent request by the NGSO MSS operators to utilize the spectrum assigned to their use for MS, as well as MSS in the United States [8]. In this situation, opposition has been fierce. This is largely because the NGSO

MSS allocations are located in one of the most valuable parts of the frequency spectrum. Not only would the NGSO MSS licensees be provided with such spectrum without competition, they would be provided it for free of charge (because it is already awarded, while the rest of the MS spectrum must be assigned through an auction process).

Another approach is that of structuring a market for gray or open spectrum. Under this proposal, proposed users of the spectrum would gain access to it based on the economic value they place on such use. Accordingly, this would be a market-based approach. Critics, however, argue that the use of a gray-spectrum regime would result in an inefficient use of the spectrum because it would not be technically based. Others argue that antennas could be developed that, if required by governments, could allow multiple services to operate without causing one another harmful interference. Gray spectrum is still an early concept and one that has not yet been widely accepted or put into use by any single country.

Relocation of existing users

This section provides a brief overview of the regulatory implications of the relocation of existing users, and it is supplemented with the discussion in Chapter 9. Relocation of existing users can occur when the government either relocates the relevant frequency band to another service or authorizes it for a competing use. It is important to note that relocation of existing users tends to be a greater problem in more developed countries with competitive markets, where many different services currently operate throughout the radiocommunications spectrum resource, and congestion is imminent. Over time, however, as more and more spectrum-based technologies are utilized for telecommunications, it is likely that almost all countries will have to face similar issues.

As discussed, a large portion of the most attractive part of the radiocommunications spectrum is already utilized. Accordingly, in quite a few recent cases, new and innovative uses, such as NGSO FSS and PCS, have sought to have existing users reallocated to different frequency bands, so they can utilize the spectrum they were operating in on an interference-free basis. This is generally based on:

1. It is impossible for both the new users and the existing users to operate without one or both suffering harmful interference.

2. The new use must utilize this spectrum in order to operate.

3. Other spectrum is available for the incumbent use to which to move and in which to operate.

In such cases, the incumbent users may be asked or request to be moved out of the band. However, it is far more likely that the proposed new user will request the regulator to move the incumbents to another frequency band.

Regulators in most cases do not like to disturb existing users of the radiocommunications spectrum resource who are operating efficiently and with customers. However, as the spectrum resource becomes increasingly congested, relocation is an increasingly attractive option for governments. This is most likely the case where the existing users may be few, the potential new use can demonstrate that it is in the public interest, and the two uses will cause harmful interference into one another.

Most regulatory regimes do not adequately address how to relocate existing users of a frequency band. However, in many cases, a government will endeavor to be as fair as possible to the existing user and to provide them with a solution that allows the continued operation of their telecommunications service. In terms of administrative fairness, this generally means that a public proceeding will be held to examine the proposed relocation before a resolution is reached. However, as was the case with regard to the FCC's moving the point-to-multipoint users in the 18-GHz band in order to enable interference-free use of that band by the NGSO FSS, the FCC simply issued an administrative order of the movement for the move based on security issues [9].

In most cases, a negotiated solution by the interested parties is discussed before any public proceeding is initiated. For example, a regulator may facilitate discussions between two parties with competing uses that may operate in the same frequency band. If such a solution cannot be agreed to, then the public proceeding will provide another opportunity for a solution to be found. Major issues that must be addressed are what the new users will pay for the relocation and to what frequency band the existing users will be relocated. If they are required to move, the relocated users do not want to incur any expenses, face an interruption in their services to their customers, or face a service degradation.

In telecommunications markets that are not competitive, however, it is likely that relocation matters will be decided outside of the public view and in less than a transparent manner. Accordingly, in these situations, if a government decision is made to relocate incumbent users, it is less likely that they will be treated as fairly as those users in countries with procedural fairness requirements. In fact, relocated spectrum users in these markets may find themselves relocated without cost reimbursements and in less than optimal portions of the spectrum for their operations.

A final note must be made with regard to the relocation of incumbent government users. A good example of this was the two frequency bands, the 1.8-GHz and the 2.8-GHz bands that were under review for use by 3G services at WRC 2000. This is generally a much harder goal for the private sector to accomplish than relocating other private-sector users. However, as the spectrum resource becomes increasingly congested, the private sector is often looking to the government-utilized frequency bands as a place in which to operate its new technologies and services. Generally, in order to obtain relocation of government services, the private-sector advocate must obtain political support and demonstrate that the existing users are not the optimal users for the frequency band (and are not operating efficiently). Often, legislative or executive branch action is required to be obtained by the spectrum advocate in order to be successful in this endeavor.

The assignment and authorization of spectrum to specific users

One of the most important duties domestic regulators have is to award the use of discrete portions of the radiocommunications spectrum to users. The assignment and authorization of specific frequency bands to a single user or multiple users is quite complex. To this end, governments seeking to assign spectrum have traditionally been guided by the following (not necessarily all with the same weight):

- *Financial viability.* Ensuring that the licensee has access to sufficient funds to ensure the buildout and operation of the proposed radiocommunications system.
- *Technical viability.* The ability of the licensee to demonstrate that the proposed system is technically viable and will be able to operate as proposed. This may be demonstrated through technical studies, through the credentials of the applicants themselves, or through other means.
- *Efficient use.* The licensee should make an efficient use of the spectrum, so that this valuable resource does not lay fallow or be used in a manner that is not technically optimal.

- *Public interest.* Each country utilizes a different public-interest standard when awarding spectrum authorizations. However, in general and as discussed previously, the government wishes to see the authorized user of the spectrum meet the public-interest standard, as defined by that country.
- *Avoidance of harmful interference between users.* That the proposed use will not cause harmful interference to other authorized users in the frequency band or in adjacent frequency bands.

When there is only one applicant for a particular assignment, the subjective method of evaluating these factors and assigning discrete spectrum to a particular user has been generally relied on and has not been seen as an issue of concern. In some cases, though, competing applications are submitted that are unable to operate in the same frequency band without the potential for harmful interference between users. As more and more assignments are subject to competing applicants, this subjective licensing process has been found to be flawed, as bias at times appears to be a factor in the decision-making process. For example, in some cases, incumbent service providers who have existing ties with the government may appear to be favored in the licensing process. Further, there was often delay in the decision-making process because of the complexity of such proceedings.

As applicants were able to demonstrate that governments sometimes made biased decisions or spent very long lead times, governments began to look to other mechanisms, such as lotteries and negotiated settlements, as an objective method of spectrum assignment when the potential for harmful interference existed. Over time, however, lotteries began to fall into disfavor because of the concern over trafficking in licenses that occurred, most notably in the United States in the 1980s. Trafficking occurred when insincere applicants received regulatory authorizations that they did not intend to use, who instead resold them to third-party service providers for, in some cases, substantial amounts of money. Negotiated settlements, although still used, are often quite contentious and do not always result in a solution. Further, like comparative hearings, they are often very time consuming and result in delay to market of new services. Accordingly, this form of resolution has been relied upon less and less in recent years.

Further, as governments have begun to recognize that large revenues can be earned from the auctioning of frequency bands used for mass communications–type uses (such as mobile telephony and paging), nonmonetary-based determinations of spectrum assignment are being pushed to the wayside. Instead, many governments believe that the auction process, where the winning bidder pays the highest amount for the spectrum, results in the most efficient use of the spectrum because of the self-interest of the bidder. For example, such an approach was used in the United Kingdom for 3G services.

Still, many governments are not entirely comfortable with straight monetary auctions of spectrum because auctions do not necessarily help governments to obtain certain concessions for licenses, such as providing services to rural locations. In these cases, governments looked to supplemental approaches, with many relying on combinations of beauty contests. As discussed next, there are problems associated with even this approach, and it is likely that another new approach will soon come into vogue, especially as the financial conditions of the telecommunications industry make auctions less financially viable.

Overview of assignment processes

This section provides a basic overview of traditional forms of assignment of the spectrum resource to individual users (e.g., a comparative license process), as well as the auction process, beauty contests, and negotiated settlements. Table 6.1 provides an example of which of these processes have been utilized by certain key countries in the assignment of regulatory authorizations to 3G service providers.

In all approaches, the government will generally have certain minimum qualification standards that the applicant will have to meet before being able to participate. These qualifications may include technical and financial threshold to ensure operational viability of proposed systems. Many countries do not rely on solely one approach to award

TABLE 6.1
Examples of 3G Licensing Mechanisms in Europe

COUNTRY	LICENSING MECHANISM
Finland	Beauty contest
Germany	Auction
Netherlands	Auction
Spain	Beauty contest
United Kingdom	Auction

authorizations; they will utilize whatever approach they believe to be most appropriate for the use that is still generally authorized.

It is important to reiterate that such approaches, however, are only utilized when there are competing applications for the same use and both cannot operate without one or both causing harmful interference to the other use. In all other cases, a straight application process is still generally utilized by most of the world's regulators.

Comparative hearings

Comparative hearings were one of the earliest methods of assigning spectrum where competing uses exist such that allowing both uses would cause harmful interference into the other. A comparative hearing is a fact-based hearing where the hearing officer or arbiter determines, based on a set standard, which applicant would more directly serve the governmental purposes of using the spectrum [10]. A good example of a use where comparative hearings are often employed is in licensing broadcast stations. An appeal process is often available if an applicant believes that they were aggrieved by the decision.

Over time, however, comparative hearings were heavily criticized because of their largely subjective nature. In some cases, outright bias by the hearing officer was alleged. Further, the delay associated with complex proceedings such as these resulted in delay to market of services. Nonetheless, in some countries, comparative hearings are still used, at least in the awarding of spectrum for some users, including for the awarding of radio station authorizations.

Negotiated solutions

Another approach to assigning spectrum to individual users is the use of a negotiated settlement. A negotiated settlement may be through formal or informal processes. In either case, the settlement will be memorialized in a formal governmental document with which the parties must comply. In cases where a consensus decision cannot be reached, the government will intervene and make a decision itself. In all cases, however, such procedures are generally long and drawn out.

In terms of an informal negotiated settlement, the government may request parties with competing uses of the same spectrum to meet over a set period of time, with or without government intervention, to try to come up with a proposed usage plan. This has been used in the United States with regard to competing satellite system plans in the same plan for both spectrum and accompanying orbital locations, such as in the Ka band.

With regard to formal negotiated settlements, the government will create a formal meeting group made up of interested persons, including government representatives, to meet and find a technical solution to the proposed issue. Such proceedings are often more drawn out because they more often than not have to comply with the administrative law requirements of that particular country. Accordingly, many formalities, such as formal notices of meetings, may need to be issued and requests for comments on proposals may be sought. This often results in proceedings that take additional time to complete. However, these formal procedures do provide a safeguard against decisions rendered by the government that are biased in favor of one party. A good example of such an approach was the negotiated rulemaking proceeding used by the FCC in the 1990s in the Ka-band proceeding, whereby the FCC attempted to force a negotiated settlement of the use of the 28-GHz band (see Table 6.2).

Both approaches have been met with mixed success. In some cases, it may be impossible for individual parties who want to utilize a limited amount of spectrum to determine and agree on the amount of spectrum and in what geographic locations that they will operate in without the direct intervention of an arbiter. This is because quite often certain portions of a frequency band may have preferred technical characteristics. Another reason may be that certain geographic areas may be more profitable to service than more rural or less populated regions. In addition, larger, more politically connected players may believe they have more to lose if they compromise in a negotiated proceeding than if they seek a decision from the regulator itself. This often puts the smaller, less well-financed players in a weaker position, as the larger players seek to force a deadlock so that a decision by the regulator will be required. Conversely, at times

Table 6.2
FCC Ka Band Plan

Frequency	27.5–28.35 GHz	28.35–28.60 GHz	28.60–29.10 GHz	29.10–29.25 GHz	29.25–30.0 GHz
Primary Service	LMDS	GSO FSS	NGSO FSS	MSS feeder links and LMDS	GSO FSS

smaller players are able to ride on the coat tails of larger, better financed players to obtain the results they are seeking.

However, even if a negotiated process is unsuccessful, some movement forward is likely to be obtained. Quite simply, by trying to negotiate a settlement, the pros and cons of each approach are fleshed out. This can provide the government decision makers with a good understanding of the ramifications of each approach as they make their spectrum decision. This was the approach the FCC took in the 28-GHz band proceeding. However, like comparative hearings, negotiated settlements are often very time-consuming.

Lotteries

Lotteries are a purely random method of the assignment of radiocommunications spectrum when the spectrum is contested [10]. Generally, to participate, the applicant submits an application and an administrative fee to participate in the lottery. In the application, the applicant may have to meet certain minimum requirements to participate, such as having a locally incorporated legal entity, having minimally qualified technical personnel, and having minimum funding.

The lottery is then held in a blind manner, with any applicant that meets the basic application requirement entered into the lottery. The winner or winners of the lottery are blindly selected. Once an applicant wins, it may or may not be subject to another regulatory fee to cover the cost of the license or other government-imposed fees. In general, such fees are nominal.

Although quick, lotteries are not perfect. A major complaint of lotteries is that they often include a number of insincere applicants. For example, during the 1980s in the United States when cellular services were first being licensed widely, many shell companies were formed solely to enter the cellular lotteries. If the shell company won the license for a particular geographic region, it was often then sold to a cellular operator for substantial funds. This form of license resale likely increased the cost of cellular services in the United States.

Of course, competitive safeguards against license trafficking can be instituted to protect against such a result. This requires the imposition of such requirements as building out a system in a certain amount of time with set coverage. Failure to meet such a requirement would result in license revocation. Other solutions include not allowing the transfer of licenses until buildout occurs.

Lotteries have fallen into disfavor as auctions took the forefront, and governments realized they could obtain revenue for the use of the spectrum by the private sector. Few, if any, countries still utilize the lottery process for the assignment of the spectrum resource, especially where there are a large number of applicants. However, it is important to note that even in situations where lotteried authorizations have been resold, it is likely that this is still cheaper than an auctioned authorization.

Auctions

As discussed, auctioning of the spectrum resource is becoming the most common form of assignment process when mass communications is involved. Currently, almost all countries utilize a form of auction process for mobile telephony services. In general, auctions are monetarily based. Generally, but not always, certain criteria must be met before an applicant can participate in auctions. Such criteria may include meeting requirements, such as:

- A demonstration of the financial ability to construct and operate a system;
- Local incorporation of a legal entity;
- A certain number of technically qualified employees;
- The ability to meet all technical requirements of the proposed use;
- A demonstrated track record.

Applicants who meet these requirements may participate in the auction process. In strict monetary auctions, the qualified applicant who bids the highest monetary amount is deemed the winner (although there are other auction methodologies, such as sequential bidding). The rationale is that this applicant, valuing the resource the highest, will make the best use of the resource, hence serving the public interest. In order to ensure this goal, however, governments also condition the final assignments and authorizations provided to the auction winner with minimum buildout conditions and terms. Such conditions and terms may include a minimum geographic service area or number of consumers to be served or a set term of years during which the system must begin its operations. Failure to meet such terms could result in a revocation of the assignment and authorization.

Not all monetary auctions, however, are based solely on money. In some cases, in order to achieve certain public-interest goals, bidding credits may be awarded to certain categories of applicants, such as women-owned businesses, small businesses, or historically disadvantaged persons. These governments believe that it is important to level the auction playing field because such groups typically have a harder time gaining access to the financing necessary to win an auction.

Further, auctions are not always based on upfront payments. In Hong Kong, the government holds auctions for 3G assignments whereby the bidding is based on a royalty payment plan. Bidders offer royalty rates, which are based on a percentage of future revenues, as opposed to the traditional up-front payments typically associated with auctions of the radiocommunications spectrum resource. Further, the Hong Kong government requires the winning bidders to make a portion of the assigned spectrum available for lease, at reasonable rates, to losing bidders or other service providers who did not participate in the bidding process.

The benefits and disadvantages of the auction process

Advocates of auctions believe that there are substantial benefits to the auction process over more traditional assignment and authorization processes for spectrum. Similarly, critics often dispel some of these benefits and add other concerns about the use of auctions.

In terms of the benefits of auctions, advocates often point to speed, transparency, preservation of the public interest, and promotion of efficient, high-value use as reasons that auctions should be utilized. Of course, many critics argue that these are not genuine reasons. Here is a summary of these and other arguments both in favor of and against the use of the auction process as a method of awarding spectrum in contested matters, and a brief discussion of the validity of these concerns.

1. *Auctions are quick.*

   Pros: Auctions potentially provide a mechanism to assign spectrum that is contested in a manner that minimizes delay and improves efficiency.

   Cons: In order to adopt the appropriate bidding rules associated with them, usually through an open public proceeding, auctions are no quicker than traditional methods of assignment to be concluded.

Reality: Once a process is established in an individual country, auctions are generally a faster form of the authorization process than comparative hearings and the like. However, it is not clear that they are any quicker than the beauty contest, lottery, or similar approaches.

2. *Auctions are transparent.*

   Pros: Straight financial auctions avoid the appearance and the reality of the government making subjective decisions that may be tainted with bias towards or against a private-sector member.

   Cons: While it is true that auctions are transparent, they are no more transparent than other nonsubjective forms of authorization, such as lotteries.

   Reality: Both sides are correct.

3. *Auctions preserve the public interest.*

   Pros: Because auction winners place the highest value on the spectrum, their assignment provision means that the public recovers the full value of the spectrum, and the spectrum is put into the use that is of the highest value to the public.

   Cons: There is no public-interest determination made with regard to auctions. Accordingly, there is no guarantee that the public interest is preserved through an auction process. Further, it can be assumed that where a licensee has paid large sums for an authorization, the subscriber rates will be higher for the particular service.

   Reality: Generally, the government, in establishing the use of the relevant frequency band, has already made a determination that a specific service will serve the public interest. However, the auction winner may be able to force certain concessions from the government, possibly negating some of the anticipated public-interest benefits.

4. *Auctions promote efficient, high-value use.*

   Pros: In theory, the person who values the spectrum the highest will obtain the spectrum, thereby helping to avoid wasteful management assignment of this scarce resource. The auction winner will have the incentive to utilize the spectrum as quickly as possible in order to earn the expected investment recovery and make a profit.

Cons: While in theory this is true, in fact, and especially in auction processes where bidding credits are not awarded, the largest, best-financed entities end up with the spectrum. These entities may be able to bear the expenses of warehousing spectrum until they determine the appropriate use.

Reality: Both sides are valid. However, if buildout rules are implemented, then the auction winner is bound to meet its obligations. If they do not, however, a time-consuming revocation proceeding may need to be held, further delaying the use of the spectrum. In addition, in times of financial downturn, service providers may outspend on auctions at the expense of the adequate buildout of their proposed systems.

5. *Auctions are illegal under international law.*

   Pros: Under the Outer Space Treaty of 1964, outer space is the province of mankind. In addition, other treaties put into question the ability of anyone to claim an ownership interest in radiocommunications spectrum. Accordingly, a government cannot auction off the right to utilize spectrum outright.

   Cons: The Outer Space Treaty of 1964 only addresses outer space. It is unclear that the radiocommunications spectrum resource was included in such a definition. Further, even if radiocommunications spectrum is included in the definition of outer space, only the outright sale, not the authorization of the use of this resource, is banned. As the auction process is not an outright sale, it is not banned and is similar to any other type of assignment procedure.

   Reality: This issue is still untested under international law. What is clear is that the auction process, if in appearance only, grants a greater right to use of the spectrum than a more traditional application process.

6. *Auctions are a good way to raise revenue for the treasury.*

   Pros: The funds from spectrum auctions are an excellent way to increase the coffers of the government treasury. Such funds are then available for other important government projects, such as tele-medicine and public health. In some cases, such funds even finance the regulator.

Cons: Looking at auctions as a means to finance either the regulator's activities or other government actions is often at odds with the mission of regulators, which is generally to promote the growth of communications services through careful management of the spectrum resource. In fact, auctions may have harmed this goal, as they drain significant resources from services providers, making it impossible for them to roll out enhanced services in a timely and cost-effective fashion, if at all.

Reality: While spectrum auctions may provide funding to the government, when vast sums are involved they most definitely jeopardize the ability of service providers to provide the services they planned. An excellent case in point is the 3G auctions held in the United Kingdom in the early 2000s. During that auction, British Telecommunications paid outrageous amounts of money for its 3G authorization. Over time, it became evident that this drain on financial resources almost bankrupted one of the world's largest service providers and resulted in British Telecommunications divesting itself of its 3G business in order to stay solvent.

An exception to auctions in the United States

An interesting caveat from the trend towards auctions occurred in the late 1990s in the United States. The satellite industry led a fierce battle within the U.S. Congress in order to ensure that it would not be subject to auctions for international satellite services. However, the auction of domestic satellite services was not similarly prohibited. In fact, the United States has used auctions to award satellite licenses in two instances—for a domestic direct broadcast satellite service and for the digital audio radio satellite service.

The fear of this community was that because satellites operating internationally must obtain service authorizations in many countries, these services could be saddled with an auction process in each country if the United States started the trend. Further, these interests also believe that having a coordinated multinational auction would likely involve a substantial investment of time and resources by multiple governments. This could raise issues of national sovereignty and access, and would undoubtedly result in a delay of the service to market. However, domestic auctions of satellite authorizations were not prohibited, as the cross-border concerns associated with international satellite systems are not present. Whether the decision by the United States to specifically prohibit by law auctions for satellite services will impact the decisions of other countries with regard to satellite auctions remains to be seen.

Beauty contests

A hybrid approach to the assignment of spectrum, the beauty contest, has both a subjective component and an auction aspect. In this regard, beauty contests generally involve a subjective application of what the applicant is willing to do (often in terms of quality of service, pricing, and geographic reach) and how much the applicant is willing to pay for the right to provide this service. The regulator then evaluates these factors to determine the winner.

Beauty contests are often criticized as having the worst of all assignment processes; they are subjective and monetarily based. However, many advocates believe that this approach is a much better approach than either of the others—it takes the best from both worlds toward the authorization of an entity that values the spectrum the highest. Many countries have seen successful completion of the assignment process for fixed wireless uses by using beauty contests, including Belgium and Denmark, both of which faced oversubscription in the number of applicants.

The implementation and enforcement of technical and operating rules

Another key role of domestic regulators is the implementation and enforcement of technical and operating rules governing spectrum usage. The importance of this role and the ability of the regulator to fulfill these functions vary greatly by country. For example, in some markets, like New Zealand, regulation, including enforcement, has traditionally been very weak. Other markets, such as Brazil, have been very strong in the areas of implementing and enforcing their technical and operational rules, issuing fines and other punitive actions against licensees that have been found to be in violation of legal requirements.

In terms of implementation of technical and operating rules, in order to ensure efficient operations, it is imperative that the regulator adopts and implements timely and relevant technical and operational rules. Failure to act may directly result in the failure to introduce new technologies and services to consumers or may allow inefficient operations to occur.

Unfortunately, the regulatory process takes time—especially in open markets where strong administrative procedure rules exist. In these markets, such rules cannot be adopted without public input. Accordingly, quite often, necessary rules changes can take well over a year to be put into place. This time delay directly impacts the operations of telecommunications service providers and network operators and may impact the business plans of telecommunications equipment manufacturers. Yet this disadvantage must be seen in light of the benefits that an open regulatory process affords in terms of fairness.

Enforcement of technical and operational rules (including the terms and conditions of assignments and authorizations) is generally of even greater importance. A weak regulator may allow anticompetitive activities to exist in the telecommunications market, allowing a particular competitor to have a leg up on others. Similarly, violations of technical rules in the spectrum arena may result in harmful interference into the operations of other service providers, including those operating in a manner that is compliant with such rules.

The regulation of secondary markets

A growing area of importance to domestic regulators, especially in light of the recent downturn in the economic condition of the telecommunications industry, is regulation of secondary markets in spectrum assignments and authorizations. A secondary market is created when governments allow licensed service providers to either sell or lease their right to use the spectrum for part of or the full license term. Although concern over trafficking in assignments and authorizations has always been a concern, the use of secondary markets has become a more visible practice, as more and more companies who hold multiple spectrum assignments, such as Winstar, wind down their existing operations and seek buyers for their assets, including these assignments. What follows is a brief overview of this area, with Chapter 8 providing an in-depth examination.

Most regulatory authorizations provide for certain limits on transferability in order to avoid the ability of companies to traffic in regulatory authorizations. Such limits may include minimal holding periods of the authorization and buildout obligations. Further, most governments require that any transfer of control of an assignment of spectrum must be approved by the government. Accordingly, there are certain safeguards against some transfers of assignments. For example, many governments' regulations prohibit speculation and trafficking in wireless telecommunications licenses, construction requirements for the licenses, and the potential for fraud in the licensing process. Governments have traditionally imposed such rules to ensure that licensees are planning to and do use the spectrum that they are authorized to operate within.

However, as the telecommunications industry becomes retrenched, many companies that may have met the initial license conditions are now forced to totally divest their regulatory authorizations. This has created a new realm of issues with which regulator must grapple.

In addition, many governments, such as the United States, Australia, and the United Kingdom, are either exploring the possibility of or already allowing entities with radiocommunications spectrum assignments to sublet these assignments to third parties. For example, in the United States, the MMDS operators currently lease a substantial amount of spectrum from the ITFS licensees. These licensees are generally nonprofit organizations.

Australia had one of the first formalized regimes in place for the reallocation and trading of radiocommunications spectrum. Under this structure, although use is allowed by third parties for compensation to the licensee, the licensee remains subject to the requirements of its authorization and assignment. To this end, the third-party use is subject to termination at any time.

Conclusion

The focus of the past few chapters has been on the broad international and domestic regulatory regime governing the use of the radiocommunications spectrum resource. During the course of this review, many issues have been raised. These include matters of choosing the best method to award spectrum when there are competing uses, regulating secondary markets for the spectrum resource, and finding the appropriate method for the relocation of existing users. In subsequent chapters, many of these issues will be revisited as the very contentious nature of the use of the spectrum resource is further explored. The issues inherent in both types of secondary use will be explored in latter portions of this book.

Endnotes

[1] See, for example, http://www.oftel.gov.uk; see also http://www.dti.gov.uk/sectors_telecomm.html (providing overview of the U.K. regulatory scheme).

[2] http://www.anatel.gov.br/home/default.asp.

[3] http://www.ntia.doc.gov.

[4] Gurreiro, Renato Naverr, "The Telecommunication Framework in Brazil," National Telecommunications Agency, 2000, available at http://www.anatel.gov.br/Tools/frame.asp?link=/english_site/lecture/palestra_navarro_abta_08_12_2000.pdf.

[5] The WTO Agreement Reference Paper available at http://www.wto.org/english/tratop_e/serv_e/telecom_e/tel23_e.htm.

[6] http://www.skystation.com.

[7] Amendment of Part 2 of the Commission's Rules to Allocate Spectrum Below 3 GHz for Mobile and Fixed Services to Support the Introduction of New Advanced Wireless Services, Including Third Generation Wireless Systems, Order, ET-258-002, September 24, 2001, available at http://ftp.fcc.gov/Bureaus/Wireless/Orders/2001/fcc01256.txt.

[8] FCC, "Public Notice, FCC Initiatives Rulemaking on Flexibility in Delivery of Mobile Satellite Services," August 9, 2001, available at http://ftp.fcc.gov/Bureaus/International/News_Releases/2001/nrin0113.html.

[9] FCC, "Public Notice, FCC Unanimously Affirms 1997 DEMS Relocation Order," ET No. 97–99, July 17, 1998, available at http://www.fcc.gov/Bureaus/International/News_Releases/1998/nrin8023.txt.

[10] Oppenheimer, Judith, "Ding, Dong the Auction's Dead," *Access Business On-Line,* July 24, 1997, available at http://icbtollfree.com/pressetc/dingdongtheauctionsdead.html (summarizing comparative hearings).

7

Solutions to Harmful Interference

Overview

This chapter explores the very crux of many of the most pivotal and contentious battles over the use of the radiocommunications spectrum resource: how governments can accommodate as many uses in given frequency band range as possible, consistent with governmental goals, while avoiding the potential for harmful interference among these operations.[1] In this case, the government will generally have to require more than mere technical coordination among different spectrum users.[2] Here, the ability of the different uses to operate in the same frequency band must be evaluated to determine the potential for harmful interference and how to best minimize that potential with minimal negative impact on operations. Accordingly, where there are multiple uses desiring to operate in the same discrete portion of the radiocommunications spectrum, in order to accommodate the uses, the government must determine whether cofrequency sharing is technically feasible and in the public interest or if another

1. This is in contrast with the scenario in which coordination is required for multiple users of the same use to operate in a discrete frequency range.
2. The international coordination of spectrum is beyond the scope of this book.

solution is more appropriate that may limit access to the spectrum to one or more of the uses.

To make such a determination, and ultimately to resolve such issues, the government, the private sector, and other interested parties often expend substantial resources, in terms of time, money, and expertise. As will be discussed, often costly regulatory proceedings are commenced, accompanied with heavy lobbying efforts of the government by the interested parties on their preferred outcome.

The focus here is to first supplement earlier discussions on why such conflicts among competing uses may arise. This chapter then focuses on certain principles that governments may implement before such conflicts develop in order to minimize their potential occurrences. Governments often provide incentives for incumbent uses and new uses to develop, utilize, and implement technologies and equipment that allow for efficient use of the radiocommunications spectrum resource, as well as to mitigate the potential for interference between different uses of the spectrum where technically feasible.

This discussion is supplemented with an overview of the approach of many governments to designate certain portions of the radiocommunications spectrum for use on an unlicensed basis. Unlicensed spectrum use has been used as a successful regulatory tool in some cases to force multiple uses to work out spectrum-sharing strategies between themselves while allowing flexibility on the types of use that can be made of the relevant frequency band.

Before focusing on the different methodologies available to resolve spectrum use debates, this chapter outlines the different types of regulatory proceedings and tools that are available to governments to implement such solutions. Following this discussion, we focus on the two primary methodologies that governments use to permit the use of a discrete portion of the spectrum resource by multiple users or uses, if they believe it serves the public interest—cofrequency sharing and frequency-band segmentation. Interestingly, the chapter points out that sometimes both methodologies, as well as relocation, may be used as a solution.

The first methodology discussed, cofrequency sharing, involves the adoption of technical and operational rules governing the use of the relevant frequency band, which results in multiple uses being able to operate in the same frequency band. The second methodology is frequency-band segmentation. Frequency-band segmentation results in a split of a single frequency band between different uses. This solution is often seen as a more severe remedy because it generally results in each use having access to less than the optimal amount of spectrum within which to operate. Further, it quite often requires costly system redesigns by operators or may result in degraded system performance or other operational pitfalls on both the incumbent and the new use.

The latter half of this chapter supplements the potential sharing approaches by focusing on a more extreme form of action the regulator can take with regard to accommodating new uses in a frequency band—the relocation of existing uses (including individual users) in order to accommodate a new use. This remedy is generally pursued when all other approaches have been explored and have been found to be unusable or when the new use is seen by the government to be of such a great benefit that the negative impact of the relocation is outweighed. Often, relocation proceedings are quite contentious and costly for both the incumbent use and the new use. However, because of the increased operational flexibility provided to the new use by obtaining clear spectrum, many advocates for new uses are willing to absorb these costs. Quite often, the only true adversaries to relocation is the incumbent use and any potential allies it can muster.

Spectrum conflict: the potential for harmful interference

As previously discussed, the radiocommunications spectrum is perceived as a scarce and valuable resource. One of the primary reasons for its scarcity is that today's technology allows access to only a limited amount of the radiocommunications spectrum resource. In some cases, this is the direct result of technology not catching up with science; today's technology is unable to overcome the operational limits inherent in some portions of the spectrum. In other situations, it is the result of a situation in which equipment manufacturers and network operators either do not want to invest the funds that are necessary to utilize the spectrum resource as efficiently as possible or believe that such costs would render the service uneconomical to provide or sell.

There are also other reasons, more practical in nature, for the appearance of the scarcity of the radiocommunications spectrum. In some cases, conflicts among incumbent uses or proposed uses may occur for purely anticompetitive reasons. An existing use that sees a new use of the same frequency band as a potential competitor may argue that a potential for harmful interference exists. This may be the case even where such a determination is not factually supported in order to hinder competition. Unfortunately, it is not always easily discernable where this is or is not the case because technical studies that support such claims can easily be developed and may be difficult to successfully refute.

In other cases, economics may play an important role in encouraging disputes between potential new and incumbent uses in the same frequency band. In such cases, conflicting uses may be able to share discrete portions of the radiocommunications spectrum but may require one or more of the uses to expend substantial funds to do so. For example, new technology may need to be developed to ensure the operation of both uses in a single frequency band. Even more common is that additional or more expensive equipment, such as additional antennas, may need to be deployed to allow cofrequency operations between uses. Such additional costs may make the incumbent or proposed use uneconomical to provide.

Even in cases where the additional expense surrounding the use of new technology is not of significant concern, there may be time lags between the development and manufacturing of appropriate technology to allow cofrequency operation of different uses. It often takes many years to develop new technologies and even more to manufacture them on a cost-effective basis. Accordingly, by the time such a technology is developed, it may be too long a time delay to allow the new use to be competitive in an increasingly fast-paced marketplace.

Despite the issue of scarcity, there is a growing demand by commercial and noncommercial users and operators for access to large portions of the radiocommunications spectrum resource. Over the past few years, a multitude of new technologies have been developed that require access to large amounts of the radiocommunications spectrum resource. For example, 3G advocates have argued that they require approximately 160 MHz of spectrum in order to successfully operate a 3G system. This is in comparison to the substantially less amount of spectrum that the earlier generations of mobile telephony systems required to operate within. Other recent new uses that have required large amounts of spectrum include the HAPS to provide broadband services and the spectrum required by some fixed wireless terrestrial systems to supplement existing infrastructure. A review of the upcoming agenda for the next WRC reveals that there are many other planned new wireless advanced communications systems that will also continue to place increased pressure on greater access to the radiocommunications spectrum [1].

Because of the technical limitations that exist on access to portions of the spectrum resource, coupled with the growing demand for access by both existing and new uses, the potential for spectrum scarcity has increased. Over the past few years, the greatest pressure has been on access to the lower frequency bands that have improved propagation characteristics. For example, in the 1- to 3-GHz range, rain fade is of less concern than it is in the upper frequency bands, so there is a greater demand for access by different users whose services would be most impacted by such a phenomenon.

One of the most notable of new uses that has been able to gain access into this frequency band range is 3G. However, the 3G advocates have found that gaining access has not been easy. Not only are such frequency bands congested, with services such as MMDS, but these existing users are often politically powerful and have been able to defend their right in some countries to operate in the frequency range to which the 3G advocates have sought access. In addition, because of the global roaming features associated with 3G services, many advocates have been attempting to obtain global harmonized spectrum for such use. Because of the sovereignty of individual countries that is associated with decisions on the allocation and assignment of the radiocommunications frequency spectrum resource, such harmonization is not easy to achieve.

Minimizing the potential for conflicts

From a pure public policy perspective and from an objective viewpoint, all users of the radiocommunications spectrum resource should have the incentive to minimize the potential for harmful interference in order to ensure the greatest access to all uses. Nonetheless, in reality, that is not always the case. Many incumbent users see new uses as unwanted competition or fear that allowing a new use into their frequency band will lead to greater expense or have other negative effects. In addition, governments may also have certain goals that may result in limiting access to the spectrum resource. For example, the government may want to promote a certain use of the spectrum even if this results in a detriment to other uses.

Assuming, however, that the goal of the government is to encourage more efficient use of the radiocommunications spectrum and, where appropriate, the successful sharing of the radiocommunications resource by multiple uses, governments may provide for certain incentives to spectrum users. Among the most successful of these methods are the following:

- *Incentives to use underutilized frequency bands.* Governments may provide incentives for industry and government system operators to utilize less congested areas of the spectrum. Such incentives may include providing the proposed entrant into underutilized spectrum with priority in the application process or with a waiver of certain regulatory and other fees associated with the use of such spectrum.
- *Incentives for the development of new technologies.* Governments may provide incentives for private-sector, government, and other spectrum users to create technologies that are efficient and allow greater access by a larger number of users and uses to the spectrum resource. This may include providing low-interest government loans or other financial assistance to encourage the development of such technologies, as well as providing preferential treatment in the authorization process. An example of such a program was the FCC's pioneer preference program developed in the early 1990s. Under this program, parties that demonstrated their responsibility for developing new spectrum-using communications services and technologies were granted preference in the FCC's licensing process [2].
- *User importance or priority.* Another methodology that some governments use is to provide priority to the use that it believes is of greater importance. For example, governments may be willing to provide priority to public-safety services, such as those provided to the police, at the expense of private-sector communications services that operate in the same frequency band.

However, while somewhat helpful, in most cases governments have found that incentives are generally not sufficient to ensure that significant portions of the radiocommunications spectrum are used on an efficient basis or in a method that allows as many uses as technically feasible. Accordingly, governments often have to impose requirements on the use of certain technologies or impose operational and technical limitations in order to achieve operational efficiency by spectrum users.

Some of the more common operational and technical requirements utilized by governments to achieve these goals include:

- *Requiring the use of efficient technology.* Regulators may adopt rules that require operators to use the most efficient technology available. Such rules may include requiring the use of equipment to provide service that has spectrum-reuse capabilities or has spread-spectrum capabilities.
- *Efficiency in assignment and licensing methods.* Governments may enact safeguards to ensure that spectrum is assigned in the most efficient manner possible. For example, a regulator may require applicants to demonstrate that their proposed use of the radiocommunications spectrum will minimize the potential for interference with other users and will be efficient. In some cases, if the licensee fails to meet these requirements, it may face fines or revocation of its operating authorization.
- *The role of standards in ensuring efficiency for equipment.* Governments may require authorized users to implement system designs that minimize interference with one another or that screen out unwanted radiocommunications, to the extent it is technically and economically possible. For example, the government may require the use of a certain technology, such as *time-division modulation access* (TDMA), by all operators in a specified frequency band.
- *Flexible use.* An increasingly attractive method to governments for avoiding the spectrum conflicts inherent in wireless services is to allow flexible uses in a specified frequency band. In such cases, the government may assign spectrum to specific users yet allow them to utilize the spectrum for a range of services (e.g., MS and FS), subject to certain minimum technical standards. Often, because of the inability to provide for technical rules that will protect all uses from harmful interference, there may be no guarantee that harmful interference will not occur between different users or uses of the subject frequency band.

Most recently, some dramatic changes have been suggested for the regulation and use of the radiocommunications spectrum resource. Most notable among these is the declared use of gray spectrum—that is, undoing any required usage requirements and allowing free use of the radiocommunications spectrum resource with access and allowable use based on the highest economic value placed on the spectrum by a user. This has put many incumbent uses of the frequency bands where this has been proposed on notice that their established rights to a very valuable resource may be short-lived.

An anomaly: unlicensed spectrum usage

As discussed earlier, in some cases, such as the flexible-use scenario, users may be willing to accept the potential for harmful interference from other users. This has traditionally been the case for the operation of devices that may have the potential for limited interference or for whom the burden of obtaining such protection through the licensing scheme is too onerous. In order to provide spectrum for such uses, governments often set aside a discrete amount of the frequency resource for such uses on an unlicensed basis, as opposed to requiring these users to obtain a formal regulatory authorization.

Often, unlicensed spectrum is used by low-power devices, such as television remotes, remote LANs, and baby-monitoring devices. While there may be no guarantee against harmful interference, the government and the users may have the incentive to try to minimize such potential. Similarly, low-power devices are not likely to cause interference alone or in ubiquity if the transmitters are not close enough to cause a collectively high signal level at any point. Accordingly, the users and the government often work to adapt the technologies that are utilized in the band or work out procedures among themselves to avoid such interference. In some cases, the government may also act to impose minor technical limitations on such operations.

There are both advantages and disadvantages to the use of unlicensed radiocommunications spectrum. These include:

Benefits

- *Unlicensed spectrum use saves time and costs associated with regulatory proceedings.* Because use of this spectrum does not require a license, the time and costs associated with this process are saved. The saved resources can be allocated to other uses, such as system development.
- *Lack of technical standards also decreases cost and time.* Most uses in the unlicensed frequency bands are faster and cheaper to deploy, as they do not have to conform to rigid technical standards.
- *Unlicensed spectrum use increases flexibility.* Often, unlicensed spectrum provides operators with the flexibility to provide services of

their own choosing. There may be some limitations, but in general a broad range of services can be provided.

- *Unlicensed spectrum use preserves the spectrum for future uses.* By not assigning the spectrum to specific users, at least portions of this spectrum may be available for future uses that may be developed by providing flexibility in use.
- *Unlicensed spectrum use promotes spectrum sharing.* By not licensing individual users, spectrum sharing promoted as an unlicensed use can operate while other users of the same frequency band remain idle.
- *Unlicensed spectrum use promotes innovation.* The use of unlicensed spectrum facilitates experimentation and innovation because the cost and effort of obtaining an authorization is not required. The technical limitations on the use of the band are generally not very stringent.
- *Unlicensed spectrum use increases mobility.* By not requiring individual licenses, this regulatory scheme may encourage mobility. This is because a user does not have to obtain an authorization for each site in which it may operate, as may be the case for some more traditional uses.
- *Etiquette rules may limit greed.* Spectrum greed by individual users or a particular use can be limited by imposing penalties if etiquette rules on use of the spectrum are not abided to by all users. An etiquette rule is a rule that is expected to be agreed to by the unlicensed user to minimize the potential of harmful interference between unlicensed users.

Drawbacks

- *Unlicensed spectrum use increases unpredictability.* There is unpredictability associated with the use of unlicensed spectrum. For example, the user generally has no rights if the government chooses to reallocate or reassign the spectrum to another use.
- *Unlicensed spectrum use offers no protection.* Rarely is there any guarantee against receiving harmful interference from other users in the same or adjacent frequency bands.

- *Unlicensed spectrum use increases inefficiency.* Unlicensed spectrum use is often inefficient because of the lack of technical requirements governing such use and the lack of common efficiency standards.
- *The potential for interference is increased.* Mutual interference between uses in unlicensed frequency bands may occur because there are few controls on transmissions.
- *Unlicensed spectrum use increases the potential for overuse.* There is little incentive for devices to conserve spectrum, so a user may overuse shared spectrum at the cost of other users—even if it means lower performance for other users. Such greed may consist of too high transmission duration, bandwidth, or power.

Because of the stark advantages and disadvantages associated with the use of unlicensed frequency bands, it is important for any user to evaluate whether the technical and operational rules governing the use of unlicensed spectrum provide it with sufficient protection to ensure its ability to provide service in accordance with its operational requirements. Failure to do so may result in the inability to provide its planned services at the level of quality that the user requires. Accordingly, many users, while finding many benefits associated with the use of unlicensed spectrum, still choose to operate under a licensing scheme in order to capture the added protections that are afforded by this regulatory scheme.

Regulatory mechanisms to adopt rules governing cofrequency sharing, frequency band segmentation, and relocation

If existing rules and regulations do not allow for a new use of the spectrum, the user of that spectrum may seek governmental intervention to allow it to operate. In such cases, if the government believes that such use a may serve the public interest, the government regulator will commence a formal or informal proceeding to examine the best method to allow such use to utilize the relevant frequency band. It should be noted that in many cases it is quite difficult for an advocate of a new use of a frequency band to convince the government that is in the public interest to even examine whether the new use should be allowed. Accordingly, advocates must be prepared for an intensive lobbying effort before a proceeding is even started. For example, both NGSO FSS and NGSO MSS advocates had to expend substantial resources, in terms of money, time, personnel, and political clout in order to obtain the agreement of the U.S. government to even examine adoption of the types of rules and regulations that would allow them to operate their systems.

Once it is decided that a proceeding or process will be used to determine whether and, if so, how to allow such use in the relevant frequency band, the government must determine what regulatory methodology it will use to make this determination. As few as 10 years ago, many governments would not hold an open proceeding on such issues. Instead, such conflicts would be resolved by the government on its own accord with little transparency. However, over the past few years, this approach has begun to fall out of favor with more and more governments, especially as global telecommunications markets have become increasingly competitive. Further, transparency has become an accepted principal in telecommunications regulation, as shown by its inclusion in the Regulatory Reference Paper of the WTO Agreement, because it provides another mechanism to ensure that all parties can participate in the regulatory process on a level playing field.

Today, more and more governments, consistent with their WTO commitments, hold transparent public proceedings on the use of the radiocommunications spectrum. As discussed next, in some cases, these proceedings may be strictly notice and comment proceedings. However, as governments strive to increase confidence in their decision-making processes, often such notice and comment proceedings are supplemented by or follow negotiations between individual parties whose aim is to come up with a compromise solution. Nonetheless, there are still instances, even in the most competitive of telecommunications markets, where the government may make a unilateral decision on spectrum use.

In general, governments employ the following types of regulatory proceedings to determine the best use of a frequency band:

- *Private sector coordinators.* Some countries, including the United States, have empowered representatives of the private sector to recommend to the subject regulator the appropriate frequencies for applicants to operate in the designated radiocommunications service. This is an interesting form of public/private spectrum management. However, some tensions exist at times over what power the private committees should have, and this use may not be appropriate in the more contentious areas.

- *Negotiations informally held, confirmed by formal governmental action.* In this case, interested users, either on their own accord or at the request of the government, meet to determine possible frequency-sharing methodologies. During these negotiations, the government may or may not be a participant. If the negotiations result in a satisfactory solution, it will be confirmed by formal governmental action, usually with public comment invited before the action is finalized.
- *Formal negotiations, confirmed by formal governmental action.* In this situation, the government convenes a formal negotiation among interested parties to determine a frequency-sharing solution. Generally in these types of formal actions, administrative procedures will have to be followed, which may slow down the pace of discussions. If a compromise solution is agreed upon, the government will still have to adopt the solution into law for it to be effective. Once again, this confirmation usually takes place through a public proceeding.
- *Formal rulemaking proceedings.* Instead of a negotiated solution, or if a negotiated solution fails to be reached or is not feasible, the government may commence a rulemaking proceeding. In this situation, generally a proposal is issued by the government and interested parties are allowed to comment. Because of the formality involved, this type of proceeding is often time consuming and expensive to participate in.
- *Unilateral actions.* In some situations, governments may dispense with procedural safeguards in order to achieve their end results. This type of action is most common in situations where national defense or other important national interests are at issue. However, some governments, especially in more closed structures, may take such an action without such a basis as a matter of course.

Even in situations where all parties have fully participated, the results may not please all interested parties. Accordingly, many times governments will face judicial and other challenges to their actions. These can drag out for years, often leaving in doubt the outcome of these proceedings for many years down the road, and leaving the operators in the frequency band with a sense of regulatory uncertainty.

Cofrequency sharing and frequency-band segmentation: an overview

This section focuses specifically on the two primary methodologies to allow multiple uses to operate in the same frequency band on the same priority basis. The first methodology is cofrequency sharing. Under this approach, governments designate a frequency band for multiple uses whose operations have the potential for harmful interference with one another. In order to minimize this potential, government regulators adopt technical and operational limitations on each use to ensure that all can operate in the same frequency band without causing harmful interference into the other use or uses.

The second methodology is frequency-band segmentation. In this situation, multiple uses may operate in discrete portions of the same frequency band. For example, a single frequency band of 500 MHz of spectrum may be segmented into two discrete portions with two different uses, with each use only allowed to operate in its own discrete portion of the spectrum. Often strict technical and operational limits are imposed on each use to ensure that the adjacent uses do not cause each other unacceptable interference.

Quite often, cofrequency sharing will be the first compromise choice of the parties involved (after having the frequency band exclusively to a single use). Unfortunately, because of the technical limitations that are often imposed as part of a cofrequency-sharing scenario, successful operation of a service may not be possible. Accordingly, band segmentation will be the end result.

Overview of cofrequency sharing and band-segmentation solutions

In almost all situations where there is a potential for harmful interference among existing and proposed uses, governments strive to avoid such potential by adopting and implementing technical and operational rules governing the use of the relevant frequency band. Unfortunately, because of the complexity of the use of the spectrum resource, such rules are not easily determined. In addition, because most service providers and network operators would like to have exclusive access to the relevant frequency band, many resist the adoption of cofrequency-sharing rules. These interests often instead argue that the competing use should not be permitted to operate in the same frequency band as it does because of the potential for harmful interference or because of the burdens being imposed on its operational abilities. Accordingly, the adoption of such rules is often time consuming and resource intensive, and still may result in less than an optimal solution for each party. However, the ability of the government to find solutions that allow the operation of multiple technologies often outweighs the ability of an advocate to have an ideal solution.

Over all, there are many different types of solutions to frequency sharing. The two major categories are those that involve technical or geographic solutions. In some situations, the government regulator may find it appropriate to adopt rules governing geographic and technical solutions. What follows is a brief description of these types of solutions.

1. *Technical solutions.* There are many types of technical solutions. These include, but are not limited to, the imposition of:
    - Emission limitations;
    - Requirements on the siting of antennas and other equipment;
    - Technical limitations on the design and operation of consumer equipment;
    - Requirements to utilize spread-spectrum or other technologies that ensure efficiency in spectrum use;
    - Requirements on the use of certain technical standards for the development and manufacturing of equipment to increase spectrum efficiency and decrease potentials for interference with other uses;
    - Requirements for coordination between operators or users of each use;
    - A channelization plan.
2. *Geographic restrictions.* Cofrequency sharing may be accomplished by ensuring that users that could potentially cause harmful interference to other authorized users are not permitted to operate in overlapping geographic areas.

The difficulties associated with reaching a cofrequency-sharing solution

Proceedings over cofrequency sharing are often very contentious. In almost all situations, interested parties would generally prefer to have sole use of the relevant frequency band, as this generally imposes the least technical, operational, and monetary requirements on either party. Accordingly, in many cofrequency-sharing proceedings, each side may try to hinder a resolution, preferring to try to obtain clear access to as much of the frequency band as possible.

Further, another general goal of the interested parties is to have as few technical and operational limits imposed on their use as possible, with the larger burden being placed on the conflicting use. This ultimately results in each side battling through its technical team to show how it does not cause interference into the other use, but the other use imposes interference into it and should be appropriately curtailed.

Despite these positions, if a cofrequency-sharing arrangement is ultimately reached, generally all uses will face more stringent technical and operational rules then they would if they operated on an exclusive basis. In fact, many successful cofrequency-sharing arrangements are concluded because the interested parties ultimately recognize that they will have to face certain constraints or risk losing access to the relevant frequency band. Accordingly, they may agree to the imposition of technical and operational requirements in order to ensure they still have access to the frequency band, even at the expense of other items of importance, such as operational flexibility. In other cases, the government will order the cofrequency sharing, forcing limitations on the operations of all uses in the band. However, this is generally not a preferred approach.

An interesting example of a successful cofrequency-sharing solution is that imposed by the FCC when it auctioned the 1.9-GHz band off for PCS. In this band, well over 4,500 point-to-point microwave paths were already operating in this same band. In order to minimize interference between operations, the FCC required each PCS operator to [3]:

- Provide interference protection to the incumbent microwave systems using the industry-accepted interference calculation methods contained in the *Telecommunications Industry Association* (TIA) *Telecommunications Systems Bulletin* (TSB) 10F;
- Distribute prior coordination notices to all incumbents within a specified coordination distance at least 30 days prior to operation;
- Submit PCS site information to the cost-sharing clearinghouses prior to operation.

In addition to these notice and protection provisions, the FCC recognized that cofrequency sharing should not be the ultimate solution. Instead, the FCC imposed a timeline during which PCS operators could ask the incumbent microwave providers to relocate to an alternate frequency band. Over time, forced relocation could occur, while the beginning time frame was on a voluntary basis with an accompanying financial compensation scheme [4].

Frequency-band segmentation

Another methodology for accommodating multiple uses in a discrete frequency range is to impose a frequency-band segmentation plan. This means that a single frequency range will be divided between multiple uses, with each obtaining a discrete portion of a larger band within which to operate its systems. Each use will be assigned a discrete portion of the frequency band to utilize with specific technical and operational rules that govern use of that portion of the relevant frequency band.

Frequency-band segmentation is often the result of failed efforts to determine a satisfactory cofrequency-sharing solution. While a band-segmentation scheme may limit the more stringent technical and operational rules typically associated with adoption of a cofrequency-sharing approach, it still has its disadvantages. For example, in many situations, the result means that there is a reduction of available spectrum to all of the uses involved This often results in less than the optimal bandwidth being available for each use, as the whole frequency band is being divided up among subuses.

Another disadvantage to this approach is that incumbent users will often face some form of relocation in order to accommodate the new use. In some situations, this may require a move of some operations to a different frequency band, while in other more extreme cases access to a portion of the spectrum may be lost entirely.

However, frequency-band segmentation, like cofrequency sharing, is often better than finding no solution. For example, by reaching agreement on a band-segmentation plan, all interested uses may be able to obtain access to at least a portion of the desired frequency band. In order to encourage such compromise, many governments will raise the potential need for relocation as a way to move things forward in a band-segmentation proceeding.

In all cases, it is necessary to adopt new rules to govern the frequency-band segmentation plan. These rules will govern such areas as technical requirements on operation and operational limits. The aim of such restrictions is to ensure that adjacent-band interference is not caused.

A major proceeding where sharing turned into band segmentation involved the FCC in the use of the 28-GHz band for NGSO FSS, GSO FSS, and LMDS [4]. Segmentation has been examined for many other uses in other frequency bands, including 3G, meteorological aids, and vehicle tracking systems. In most cases, however, it is very controversial and a difficult issue to resolve.

Relocation of existing users

Overview

As the radiocommunications spectrum resource has increased in congestion over the past few years, there has been a stronger movement towards the relocation of incumbent use of the spectrum to different frequency bands to allow new services or uses to operate in that frequency band. Relocation, also known as refarming of the radiocommunications spectrum, can be defined as:

> The cancellation of current allocations and/or assignments in a particular range or ranges of frequencies so that new users of the spectrum can operate in that band.

Relocation is the term that is generally used when such an approach is initiated in response to a private-sector or other request. *Refarming* is the term generally used when the government determines, as part of a broader review, that the use of spectrum should be changed from one use to another. As more innovative uses enter the marketplace, it is becoming more common for governments to engage in self-initiated refarming of the spectrum resource [5].

Accordingly, relocation or refarming mainly occurs because of a specified need or as part of an overall government review. In terms of governmental review, some governments, such as Australia, go through a regular review process to determine whether the radiocommunications spectrum resource is being utilized in a manner that serves the public interest or if changes should be made. In some cases, this may result in the relocation of existing uses from one frequency band to another.

In other cases, there may be a specified need being addressed. For example, a new use may be proposed and may request access to a specific frequency band. As discussed, relocation is not a remedy that is pursued or utilized in every instance. In general, if suitable alternate spectrum is available for the new use or a sharing scenario is possible, that is generally the preferred solution, at least to the government. Moving existing users almost always has associated political costs. However, in situations where there is no other possible solution, or where the government believes the new use has the potential for vast benefits, relocation may occur.

The need for comparable spectrum and reasonable compensation for relocated uses

Perhaps one of the most daunting challenges facing all parties involved in a relocation proceeding is the ability to find available spectrum to move the incumbent use to. If there are no or only a few users currently operating in the relevant frequency band, this situation is generally fairly easy to resolve. The terms of such relocation are critical issues both for the new users of the spectrum, and for the incumbents who must involuntarily relocate.

Accordingly, in the case of many incumbent users in the frequency band, such relocation is often quite complex for several reasons. First, a substantial portion of the most attractive portions of the spectrum resource is already congested. Second, it may be difficult to find spectrum that has the appropriate technical characteristics for the incumbent use. Third, there may be hostilities to moving by the incumbent users and their customers because it may involve inconvenience, potentially severe service disruptions, and even degraded service. Finally, such relocation is often quite financially costly. Such costs can include the need to build a redundant network to handle the transition and the need for different or upgraded equipment to operate in the new frequency band.

In order to address this problem, in almost all cases, relocated users of the spectrum resource request comparable spectrum to be identified and made available for their use and adequate compensation. With regard to the ability of comparable spectrum, this is often a demand before any talks of relocation are even made. For example, in the 3G debate in the United States, the DOD specifically asked that any efforts to transfer the use of the 1.8-GHz band to private industry be delayed "until truly comparable spectrum is identified and made available" [6].

However, finding comparable spectrum is often difficult. Initially, there are the congestion and technical issues associated with finding such spectrum. This is often supplemented by the fact that the incumbent that is being relocated to the new spectrum may be hostile to the relocation. Accordingly, the incumbent use may claim that certain bands are not comparable in order to try and forestall the relocation or even possibly to obtain further concessions from the new users. Further, in some cases, there are existing uses in the frequency band to which the incumbent is being relocated. Accordingly, technical and operational rules may need to be implemented to allow cofrequency sharing among the uses, and, in severe cases, relocation of these incumbent uses may also be required.

Another major issue is whether relocated uses are entitled to compensation. This raises the issue of who will pay for those costs. Some new uses have claimed in the past that they should not have to pay for relocation. A good example of this is the U.S. international satellite industry. Some of these companies have argued that because they operate on a global basis, they could not afford to pay for relocation. This is because it would mean that they would have to pay for relocation on a global basis. This argument has been found to be flawed. This was a major issue in the FCC's 2-GHz proceeding, where ICO, a global MSS provider, argued this. In fact, the EU participated and claimed that the imposition of such relocation costs would constitute an unfair trade barrier [7]. The FCC ultimately dismissed this argument.

A further issue is how to determine the real costs of relocation. Such costs must be adequate and accurate. Failure to require this would impose an unfair burden on the new use. Most governments require the new entrant to pay the incumbent for:

1. The actual costs of relocation to a *comparable* location;
2. Completion of all activities necessary to begin operation;
3. Construction and testing of any new equipment and associated costs.

Governments, however, often require the incumbent to mitigate costs of the relocation. For example, it would not be equitable for a new entrant to pay for an equipment upgrade of an incumbent if it is not directly related to the relocation.

Over all, spectrum relocation and refarming are still quite controversial regulatory remedies. These tools tend to be successful for all interests involved only when the government can obtain a consensus among the parties on the relocation, the comparable spectrum to which the incumbent is being moved, the time scale, and the compensation. When such arrangements are not mutually agreed to, albeit grudgingly, the government may find itself caught in a myriad of future challenges on the regulatory action.

Endnotes

[1] ITU Agenda for the World Radicommunication Conference (WRC-03), July 26, 2000, available at http://www.info.gov.hk/itbb/english/speech/pr14022001.htm.

[2] See http://www.fcc.gov/oet/faqs/pioneerfaqs.html. This program was eliminated in the late 1990s.

[3] See http://www.comsearch.com/pcs_cellular/design4.jsp.

[4] FCC's 28 GHz Rulemaking Proceeding.

[5] See http://www.teledotcom.com/article/TEL20000825S0009.

[6] Leopold, George, "DOD Returns Fire on Spectrum Relocation," *EE Times*, August 31, 2001.

[7] ICO Press Release Commenting on FCC's Relocation Order, July 5, 2000, available at http://www.ico.com/press/releases/200007/000707.htm.

8

Secondary Markets for Spectrum

The increasing use of secondary markets

As previously discussed, the regulatory structure governing the radiocommunications spectrum resource must be able to adapt to evolving technology, changing consumer and economic markets, and the overall needs of the public and the government. Failure to ensure that the radiocommunications spectrum regulatory regime keeps pace with changing needs and requirements will, over time, result in the inefficient use of the radiocommunications spectrum resource and will not adequately serve the public interest.

Due to the competitiveness of today's marketplace and the increasing congestion of the radiocommunications resource, private industry, the government, and regulators have been looking to the creation of increasingly flexible methods to provide more efficient access to the radiocommunications spectrum for the provision of telecommunications services. This, at times, has been accomplished by allowing some classes of users to bypass the traditional rules of the regulatory arena. Instead, nontraditional mechanisms have been utilized. Governments have instituted many nontraditional methods, including management agreements, joint marketing agreements, and the resale of services.

These methodologies have provided some flexibility to spectrum users, but they have not been sufficient to ensure the most efficient use of the spectrum resource because in many situations strict limits are placed on such arrangements by government regulations. Over the past few years, in order to address such problems, private industry argued for even more increased flexibility. One of the major approaches being advocated is the adoption of the innovative concept of the authorization of secondary markets for spectrum for some frequency bands and services [1]. In essence, a secondary spectrum market expressly permits authorized users of the spectrum to allow access to this spectrum in its entirety or in part by third parties for the deployment of their own services. In many cases, such a regime is crafted so that only minimal regulations are imposed on the secondary spectrum use.

Advocates argue that allowing the use of spectrum through secondary markets is particularly appropriate, as the ever-rising demand for the use of the spectrum resource by an increasing number of operators and users, especially in the lower frequency bands, has created a shortage of available spectrum [2]. By allowing the creation of secondary markets, there may be increased use of the spectrum that is already authorized but may be underutilized or not utilized efficiently. Further, advocates believe that such spectrum may be able to be utilized for new services or uses or may be able to provide services to rural or remote locations. In addition, in some cases, it has been shown that the secondary spectrum market can be used to assist existing operators in other frequency bands, so that they can make use of additional spectrum where they may not otherwise have sufficient capacity for the service or use.

Such arguments have provided the basis for many countries to begin to allow the creation of secondary spectrum markets, at least for some frequency bands. The remainder of this chapter discusses the advantages and disadvantages of the use of secondary spectrum markets, types of secondary spectrum markets, and the criteria that are often utilized in creating a regulatory regime governing secondary spectrum markets.

Advantages and disadvantages to the use of secondary spectrum markets

It is not always readily apparent whether new regulatory solutions to spectrum use are beneficial to all or even the majority of interests. The same is true with the introduction of secondary spectrum markets. Accordingly, it is important that there is a full understanding by all interested parties of both the disadvantages and the advantage to the creation of secondary spectrum markets before crafting such a regime. Some of the major advantages to the introduction and use of such a regime include the following:

- It encourages efficiency in the use of the spectrum resource.
- It increases the potential revenue for primary spectrum license holder by allowing the license holder to lease unused portions of the spectrum that it has the right to utilize.
- It creates a method for the secondary market user to offer a service or to expand its existing service.
- It spurs market opportunities for other related sectors, such as telecommunications equipment manufacturers, by opening up more of the spectrum resource for additional uses.
- It increases access to existing and new services for consumers.
- It maximizes access to and use of the radiocommunications spectrum by multiple users.
- It promotes increased competition in the telecommunications marketplace by encouraging full use of the spectrum resource.
- It provides flexibility to operators of spectrum-based telecommunications services to respond quickly and efficiently to changing technological and marketplace needs, often without the interference of the government or with only a limited amount of interference.
- Authorizing the ability to transfer a spectrum license increases the value of the relevant portion of the spectrum.

Some of the major disadvantages include the following:

- The ability of the licensee to resell or lease rights to the relevant portion of the spectrum encourages speculation in the spectrum resource.
- The regulator may face enforcement problems because the licensee may resell its rights to the spectrum to a third party—over whom the regulator may not have jurisdiction or may not be able to adequately police.

- The temporary transfer of spectrum to a third party may not always serve the public interest. Ensuring that the public interest is met may be difficult through a scheme that does not provide a direct nexus between the spectrum user and the regulator.
- Spectrum users may be encouraged to hold onto fallow spectrum, as opposed to making immediate use of it, because it may increase in value at a later date. At that point, the user may be able to resell the use of its spectrum to others through such mechanisms as a secondary market.
- Spectrum users may be encouraged to hold onto fallow spectrum, as opposed to making use of it, in an effort to hinder competition. In this manner, users can delay access to key pieces of the spectrum to their competitors.
- It may result in unexpected financial windfall gains to the primary license holder if the spectrum is priced below value and is resold by the license holder subsequently at a higher rate.
- It may result in a single company being able to control a significant amount of spectrum.
- It may increase costs in the marketplace for spectrum-based services because of the real potential for the increase in price when the spectrum is resold on the secondary market.
- It may limit the entities that have access to the spectrum by not providing the government with a mechanism to ensure that spectrum users are diverse.
- It may create difficulties in how the government is able to impose defense and other national security concerns on third-party users of the relevant portion of the spectrum.

Governments must consider these advantages and disadvantages as they craft their regulatory regimes. A government may want to reconsider the use of such a scheme if it will have certain effects that may be adverse to the government's policies. In other cases, by understanding the impact of utilizing a secondary spectrum market for the use of the spectrum resource, the government may find it necessary to impose certain requirements on users in order to ensure its policy goals are met.

Similarly, before they choose to utilize spectrum that is subject to a secondary market regime, private parties must understand these considerations, as the existence of such a scheme may directly impact a company's business plan. For example, a company that is launching a new type of radiocommunications use may be more willing to spend a large sum in a spectrum auction if it knows that it will be able to allow other parties to use its spectrum for a fee if its own use fails or takes a extra time to be ready for service.

Further, many other interests need to be aware of the impact of the creation of a secondary spectrum market. For example, consumer groups may be leery of such a regime because of the potential for higher prices associated with services operated in spectrum that is obtained on a secondary market. Further, equipment manufacturers may find the creation of secondary markets attractive because it may allow them to sell more equipment to these users of the spectrum resource.

Types of secondary spectrum market regimes

Regardless of whether a secondary spectrum regime is crafted, there is always a primary market for the spectrum resource. This is represented by spectrum that has been assigned and licensed for use by one of the many processes discussed earlier in this book. Such processes may include lotteries, auctions, and straight application procedures. The secondary market occurs only after the primary distribution of the discrete portion of the spectrum. This spectrum is then authorized to be exchanged, traded, or leased to a third party, hence creating the secondary market. In some cases, the secondary market may be facilitated through brokers, dealers, or other intermediaries between the primary and secondary spectrum users. Use of such intermediaries may be voluntary or required by the government regime governing the secondary market.

A regulator can structure the secondary market regimes in several main ways. In some cases, there may be a lease of the spectrum to a third party for the entire term of the license held by the primary spectrum user. For example, a licensee for a mobile telephony license may not have the capital to build and operate a system, at least initially. In this case, it may make financial sense for the licensee to lease the spectrum to a third party during its license term. However, in such a situation, the regulator may be concerned about the licensee having no intent to build out its system, according to the terms of the authorization. Accordingly, in some cases, regulators may not allow a full authorization term to be leased to a third party by the primary licensee. In other cases, such a deal may be allowed, but with certain limitations placed on either or both the primary and secondary spectrum user.

A variation of this approach is the concept of the *lease* or *resale* of the primary spectrum on a temporary basis to meet short- or medium-term demand for a particular service. For example, an entity holding a regulatory authorization to provide fixed wireless access service, in anticipation of its own growth, may lease spectrum to another entity to allow the latter to meet some increase in demand. Such an increase may be caused by a temporary event, such as a large sporting event or important news event.

Another possible approach to the creation of a regime governing secondary spectrum markets is to only sublease a portion of the frequencies held by the primary spectrum holder or a limited geographic area. Such an approach may occur where a paging operator, for instance, is only able to build out in a portion of the service areas it has authority to serve in the immediate future. In this case, the primary license holder may lease out spectrum to a third party that is a discrete geographic location, such as a specific city.

Another possibility is the exchange of spectrum between two license holders. For example, it may be attractive for two authorized service providers, with access to different frequency bands or different geographic areas, to exchange spectrum so that they could both supplement their systems in a manner that they believe is rational.

In all cases, regulators have significant power to shape the secondary market, if they choose to exercise such power. A major issue that the regulator must determine is the degree to which allowing the creation of a secondary market can deliver efficiency of spectrum use, compared to the complexity it adds to the spectrum regulatory regime.

New Zealand: an overview

In order to fully understand the utility of the secondary spectrum market, it is interesting to look at the earliest use of the concept of secondary spectrum markets by the New Zealand government. However, it is important to note that although the New Zealand government has encouraged the creation of secondary spectrum markets, it has not widely utilized this approach.

In 1989, New Zealand became the first country to allow the trading of spectrum rights. It is unsurprising that New Zealand was a groundbreaker in this area, as it was the first market to truly adopt a laissez-faire approach to regulation for all telecommunications services. Under the New Zealand approach to regulation, the government permitted the market to dictate how its market for telecommunications services would be shaped. Accordingly, when New Zealand began to explore an innovative regulatory regime for the use of the radiocommunications spectrum, the government sought to adopt an approach that would move away from a traditional process to a more market-based approach to spectrum use.

Specifically, the New Zealand regime created three different types of property rights for users of the spectrum. These are [3]:

- *Management rights.* Within certain interference limits, the authorized user is granted an exclusive right to the management of a nationwide band of frequencies for a period of up to 20 years. Within this band, the authorized user can issue sublicenses to third parties (through a variety of mechanisms, including lease of the spectrum).
- *License rights.* The license allows use within a set geographic area. The licensee can use the spectrum for any service, as long as interference constraints are met.
- *Apparatus licenses.* Where management rights have not been granted, nontradable licenses exist for the use of discrete portions of the radiocommunications spectrum.

Under this regime, the New Zealand government has created the potential for a secondary spectrum market in the portions of the spectrum where management rights have been authorized but not in the other cases.

In general, management rights to the spectrum are granted in spectrum awarded through an auction process. Following the initial grant, the relevant portion of the spectrum resource can be freely traded by the initial licensee. It is up to each licensee, often know as the *spectrum manager*, to determine whether they want to trade their spectrum. The only restrictions are technical ones on interference and those governing competitive issues as set forth in the New Zealand competition law.

Despite the creation of this regime, spectrum trading has not caught on widely in New Zealand. Some of the reasons that have been attributed for this include [3]:

- Concern about the adequacy of competitive safeguards;
- That the primary assignment is market based, making the secondary market less attractive;
- That the spectrum market has become less vibrant, as evidenced by the small interest in the recent 3G auctions.

Creating a regime governing secondary markets for spectrum

Any government that is creating a regulatory regime governing secondary spectrum markets must determine which criteria and rules governing the market it wants to impose, if any. The purpose of these rules can be multifold, and include protecting against anticompetitive conduct by spectrum users and ensuring universal access to the spectrum for users. Such criteria and rules typically include:

- *A limit on the time or scope of the secondary use.* As discussed earlier, in some cases, it may make sense to impose limitations on the time or scope of the secondary use to discourage spectrum speculation.
- *The licensee retains responsibility for compliance with regulatory obligations.* In most cases, it makes sense for the government to ensure that the primary license holder retains responsibility for compliance with all regulatory obligations. In this manner, the regulator has a direct recourse and does not have to rely on obtaining jurisdiction over the secondary spectrum user in order to address issues of concern.
- *Secondary users are responsible for compliance with service, technical, and operational requirements.* Even if the licensee retains responsibility for compliance with regulatory obligations, the regulator may also want to make sure that it has a means to ensure the secondary spectrum user complies with the service, technical, and operational requirements of the regulator. The regulator may be able to do this through the enforcement of fines or other penalties or, in extreme circumstances, termination of the secondary use.
- *Commercial disputes.* The regulator should ensure that there is a mechanism contained in the agreements between the primary and

secondary spectrum users to resolve disputes. In general, commercial issues should have to be resolved between the parties, although the government may want to have notice of the dispute and the resolution.

- *Construction milestones or other buildout/service requirements.* The issue remains whether the primary licensee should be able to use the actions of the secondary spectrum user to meet its construction milestones or other buildout/service requirements. In order to encourage spectrum utilization, it generally makes sense for the primary user to be allowed to do this.
- *Flexible use of spectrum.* In an increasing number of cases, as discussed earlier, governments are beginning to establish flexible use of spectrum, so that multiple uses of a single frequency band can all operate. Of course, the regulator must weigh the creation of such approach against its public-interest goals in establishing more rigid use policies, especially in the case of secondary spectrum markets where the regulator is often more removed from the its traditional oversight function.
- *Regulatory fees.* Another issue is whether the government should impose additional regulatory fees on secondary spectrum users. In general, it may be more efficient to rely on the primary licensee to handle these, and for this to be resolved between the secondary and primary license holders on a commercial basis. In other cases, governments may wish to charge both users directly for their right of use of the spectrum resource.
- *Clearly defined rules and regulations governing secondary spectrum markets.* In all cases, it is imperative that the government creates a transparent process with clear rules and regulations governing secondary spectrum markets. This will help build confidence in the process and likely increase the efficiency of the operation of the secondary market.
- *Methodologies to bringing primary and secondary buyers together.* In order for secondary spectrum markets to be successful, there must be an efficient methodology for bringing buyers and sellers together. In most markets, there is no standard method for the secondary spectrum market to operate. Regulators may look at developing and encouraging this process. Methodologies they may use include

keeping an easily accessible database of primary license holders who are interested in offering all or a portion of their spectrum rights to a secondary user or encouraging private brokers to pick up this function or other similar services.

- *Protection from harmful interference for both primary and secondary users.* The government may want to ensure that some form of protection from harmful interference is provided to secondary users. Such protection should generally be the same as the primary license holder is granted under the relevant rules and policies of the telecommunications regime, such as emission limits.
- *Harmonizing operational rules for similar services to ensure broad access.* In order to allow operators to gain access to adjacent frequency bands, it may make sense to harmonize operational rules for similar services to ensure broad access.
- *A determination of whether secondary markets should be created only for spectrum that is auctioned.* It may appropriate for some markets to limit the secondary markets to spectrum that is auctioned. This is because the creation of a secondary market means that the regulator is allowing the primary licensee to make money on the spectrum. It may not be appropriate policy to allow a licensee that received free use of the spectrum to profit monetarily if they did not pay the government a fee for such use.
- *Spectrum caps on ownership.* Another consideration is the establishment and enforcement of rules to control the amount of spectrum that may be acquired by an individual service provider. This mechanism will serve to prevent a service provider from hording spectrum or otherwise negatively impact the spectrum market.
- *Setting aside frequency bands that should not be subject to a secondary market regime.* In some cases, a government may make a determination that certain portions of the radiocommunications resource should not be subject to secondary spectrum markets. This may be the case when the spectrum is used for a critical public safety use, such as aeronautical communications.

Of course, each government regulator must examine which of these criteria or rules are appropriate for their own market. For example, if a government chooses to limit the spectrum to which it institutes a secondary spectrum market, it may look at the most lucrative spectrum that is likely to lay fallow. A good example of such spectrum is the 3G licenses that were issued in Europe, Japan, and other places. In many cases, because of the delays associated with deployment, much of this spectrum may remain unused for a number of years. In addition, because of the debt many of these service providers incurred to obtain the 3G licenses, many of them are facing similar challenges to even being able to deploy their services. Secondary spectrum markets may be the solution to help these companies while providing more choices, services, and competition for consumers.

However, to ensure that there is not pure speculation on such spectrum, the government regulator may want to impose certain rules on the spectrum use, including buildout requirements. Further, to meet other goals, the government may also want to look at the imposition of additional rules. However, any rules the government wants to impose must be balanced against the imposition of such rules and the resulting limits on flexibility, and against the ability of the user to make what it believes is the most efficient use of the relevant spectrum.

Conclusion

It is clear that over the next few years, the use of secondary spectrum markets will likely to continue to increase. For many service providers, especially in markets where auctions are utilized as an assignment process, the use of secondary spectrum markets makes good economic sense. Specifically, secondary markets allows these service providers to retain for their own use the amount of spectrum that they need to make immediate use, while also allowing them to lease the right to use excess spectrum to third parties. In addition, this approach also provides a mechanism to encourage the full and efficient utilization of the spectrum resource.

Endnotes

[1] See, for example, "ITA Supports Secondary Markets," ITA Press Release, March 13, 2001, available at http://www.ita-relay.com/advocacy/adv-headline15.htm.

[2] See, for example, Hazlett, Thomas W.,"The Spectrum Allocation System," *National Press Club,* November 2, 2001, available at http://www.aei.org/sp/sphazlett011102.htm; but see Rifken, Jeremy, "May Day, May Day,"

Guardian Unlimited, April 28, 2001, available at http://www.guardian.co.uk/Archive/Article/0,4273,4177269,00.html.

[3] van Caspel, Marloes, "Spectrum Trading: Increasing the Efficiency of Spectrum Usage, Analysis," available at http://www.analysys.com/default.asp?Mode=article&iLeftArticle=992.

9

Impact of the Telecommunications Financial Crisis

Since the start of the twenty-first century, it has become evident that the telecommunications boom that followed the initiation and implementation of widespread competition in the global telecommunications market has come to a dramatic and abrupt end. The anticipated proliferation of new telecommunications companies providing an ever-increasing supply of telecommunications services around the globe has failed to materialize. Instead of the robust telecommunications market that many analysts predicted, today even many of the giants of the telecommunications industry are clinging on to their own survival, with several industry leaders going out of business. For example, as of mid-2002, WorldCom, Winstar, FLAG, XO Communications, Global Crossing, and KPN Qwest all had entered into the protection afforded by entering into bankruptcy, either through a liquidation or reorganization. The impact of these bankruptcies are far reaching, directly impacting customers, vendors, and other competitors in the marketplace.

Many experts are predicting that these recent events may just be the start of a continuing downward spiral in the telecommunications industry that will continue over the next couple of years. If correct, this leaves a much-changed telecommunications landscape in place, with perhaps a more consolidated industry [1]. However, it appears that over time, the telecommunications sector will rebound and continue to deploy innovative communications services, especially in the wireless arena [2].

In response to this economic turmoil, many influential wireline and wireless telecommunications service providers have begun to rethink and retrench their business plans, including France Telecom, Nextel, Nortel Networks, and Lucent. Many of the planned product and geographic expansions have been either stopped or significantly curtailed. In many cases, lines of business once believed to be profitable have been spun off, substantially reduced, or closed down in an effort to reduce costs and streamline businesses. For example, British Telecommunications recently spun off into a separate entity its mobile telecommunications business, including the 3G business. Other providers of 3G mobile services have significantly curtailed their business plans, often calling into question when deployment of such systems will occur.

As the telecommunications industry continues to reshape itself in response to the economical troubles it has been facing, it is likely that a more streamlined environment will be developed. While the number of players in the industry is likely to drop, equally important is that the amount of financial resources that individual telecommunications companies have to operate, supplement, and expand its operations is likely to decrease. It is unlikely that the financial community will continue to support inefficient operations or finance the growth of a company into new areas without a carefully scrutinized business case being developed.

Because of this changing business environment, the wireless industry will have to overhaul its methods of operations. Specifically, one area that needs to be revisited is how spectrum is allocated and put into use both globally and domestically. First, it is likely that certain uneconomical uses or unproven uses may not be able to continue existing operations or seek to expand into new frequency bands. Accordingly, there is likely to be a retrenchment in the provision of wireless services to the public.

Second, where there was vast competition for access to the same range of frequencies by competing service providers, it is likely that some of these players will no longer exist over time, lessening congestion. This decrease will also result in diminished financial windfalls to governments that are typically associated with the auction of the radiocommunications spectrum. As a result, governments may need to revisit how spectrum is allocated and authorized for use in order to ensure that its goals are being met.

Specifically, with less competition for access to the spectrum resource, and with less financial resources on the part of applicants, governments may want to ensure that the applicants that do obtain access to the spectrum put it into use. An example of such an innovative approach is the use of gray spectrum discussed earlier.

A third likely impact is that companies seeking access to the spectrum will be curtailed in their ability to expend funds to fight the types of protracted battles that have been necessary in recent years to obtain access to new spectrum, including the relocation of existing users. Accordingly, it may be that more and more uses are more amenable to developing sharing strategies between competing uses where technically possible and not cost prohibitive.

Key reasons for the telecommunications financial meltdown

In order to fully understand the future of the radiocommunications spectrum playing field in light of the recent financial meltdown, it is important to understand the forces that have led up to this result. Up until the start of the twenty first century, it appeared that there was no clear end in sight to the increasing perceived profitability and success of the global telecommunications industry. This was true for the wireline market, and the future was predicted to be even brighter for the wireless industry, especially for broadband services such as 3G [3].

However, such optimism was ill placed. Almost all players in the telecommunications market have faced a dramatic downturn in their financial position, even telecommunications giants such as AT&T and Deutshe Telekom. In hindsight, it is clear that the telecommunications industry, including players involved in the wireless market, engaged in many practices that have assisted in the downward spiral of the industry. These practices include:

- An over buildout of capacity by telecommunications service providers, based on faulty assumptions in forecasting the market demand of existing and new services. This overbuild may be attributed in part to the increase in capacity in anticipation of Y2K.
- In a few cases, poor accounting practices led to poor economic planning or, in some cases, alleged illegal activity in accounting for costs and revenues.

- Based in part on faulty forecast assumptions, many companies expended too large an amount of funds on the fixed costs associated with their network.
- A delay in the development of equipment necessary for new services to be deployed, such as was the case for 3G. Such equipment delays directly impacted the ability of service providers to deploy their systems and capture much-needed and anticipated revenue.
- Longer-than-anticipated payback periods to meet profitability.

Over time, other practices that have assisted in the financial downturn of what at least some analysts labeled an industry that would transcend economic problems will likely emerge and be identified.

Impact of the telecommunications meltdown

In the long term, it is likely that the telecommunications industry will rebuild itself. In order to understand this rebuilding, it is imperative to first understand the impacts of the telecommunications meltdown. These impacts include:

- *Customer migrations.* As networks are turned off or the provision of service to specific geographic sites or by product type is ceased, customers have to migrate to the networks of other providers. In addition, some customers may choose to migrate on their anticipation of such actions occurring. Further, some migrations may occur by customers being lured away by providers who are willing to provide services at cut rates for this purpose. Unfortunately for business customers, the ability to change providers quickly is less of a possibility, as they generally enter into long-term contracts with service providers in order to obtain reduced rates.
- *Decrease in quality of service.* As the financial problems of the industry continue, it is likely that service providers will continue to cut back on costs, such as network builds, upgrades, and personnel. This means that service quality may begin to suffer, as the resources required to keep up the network will not be fully expended. In cases where quality suffers, general consumers can usually change service providers with ease. However, high-end users, who generally enter

into long-term contracts to obtain cheaper rates, may be locked into long-term contracts and have to suffer the impact of such service degradation.[1] While some relief may be available in the form of service level agreements, this may not be a sufficient remedy, especially for large corporate or government users that are more dependent on quality communications.

- *Vendors may not be paid.* In the case of telecommunications service providers who are short on cash, or are stayed by the bankruptcy court from paying vendors certain past-due amounts, vendors that supply a variety of items to telecommunications companies, including network equipment, office supplies, telecommunications services, and other business support may not be paid or may be paid only following a long delay. This directly impacts other sectors of the telecommunications industry, as well as other industries.
- *Possibility for less outsourcing.* Over the past decade, it became increasingly attractive for large corporations to outsource their internal telecommunications markets to telecommunications service providers. This was generally seen as more cost-effective than continually building out and managing their own networks. However, as the major providers that supported these outsourcing contracts are either retrenched or go out of business, it is likely that many companies will rethink this outsourcing strategy and deploy and manage their own telecommunications networks. This may largely be a reaction from the recent telecommunications downfall, where large telecommunications customers may feel more secure building and operating their own network, in light of recent network shutdowns, so that they are not negatively impacted by such events in the future.
- *The creation of debt-free competitors.* Many experts believe that the Chapter 11 bankruptcy process that is available in the United States will provide companies that emerge from this process with a competitive advantage over other service providers. Companies

1. Business customers generally can obtain some financial compensation through the service-level agreements into which they have entered with telecommunications service providers, but these rarely compensate the consumer for the actual impact on them, such as lost business due to the inability to communicate with other branches when there is a network outage.

emerging from Chapter 11 have portions of their debt wiped clean as part of the Chapter 11 reorganization process. This is a U.S. remedy, as reorganization is not a recognized concept under the bankruptcy laws of most other countries.

- *The sale of bankrupt assets at bargain rates.* Another situation that has already occurred is that assets or whole companies that are bankrupt or in dire financial trouble may be sold at bottom-basement rates. Such was the case of Iridium, which was in bankruptcy and ultimately sold for a very low price.
- *Decrease in competition.* Over the past few years, numerous service providers around the globe have ceased operations because of financial concerns. This is likely to continue over the next couple of years as the financial downturn continues to hit the telecommunications industry. In addition, with financial lenders imposing more stringent financial requirements, it is likely that many new competitors will have a hard time finding the capital to support their commencement of operations and existing providers will have difficulty in securing money to finance new or expanded operations. This means that there will likely be less competition in developed telecommunications markets, and there will be less development of competitive markets in less developed markets. Further, companies are likely to have less access to capital to invest in new technologies and services, further reducing competition.
- *A tarnished industry reputation and loss of consumer confidence.* Another direct result is that because of the recent downturn and the subsequent fraud that was found to be performed by several major telecommunications providers, the industry has a tarnished reputation. This directly impacts the faith of consumers, vendors, and stockholders in the industry. In addition, governments around the world may be more skeptical of this industry, which will likely result in a reevaluation of regulation for this industry.

The rebuilding of an industry

One thing is quite certain, however—the telecommunications industry will continue to exist. Its services are critical to the lives of many people and the direct functioning of business and government worldwide. However, through the rebuilding period, the following events will likely occur:

- There will be a retrenchment period, at least for the next few years, resulting in the reduction of the number of players in the industry. Some of the existing players will go out of business, while others will consolidate to take advantage of established customer bases and networks of existing service providers. Some well-financed new niche players may also enter in the market, especially where they are able to obtain existing assets at very reasonable prices.
- Resources for expansion by the telecommunications industry into new geographic areas or product lines will be scarcer, meaning companies will have a harder standard to meet in order to justify such expansion with the finance community.
- Funding in general for the telecommunications industry will be harder to obtain, and it is likely that the financial industry will impose stricter scrutiny on such investments. However, cash-flush telecommunications industry survivors may be well positioned to finance new ventures.
- Because of the stricter financial scrutiny, telecommunications services that are generally less capital intensive with large customer demand, such as wireless services, will become increasingly attractive to deploy.
- The development of new technologies may be impacted, as funds for noncore services become scarcer. However, new technologies may be the key to the future success of companies, especially in the wireless industry, as they can increase efficiency of operations.
- Deployment of new products and technologies and expansion of network reach will in general slow dramatically unless there is a cost-based rationale, such as increasing efficiency of spectrum utilization.
- Telecommunications service providers may face increased government scrutiny in their operations through traditional telecommunications regulators and other government bodies.
- Consumers in the future will be more likely to scrutinize the financial health of their telecommunications service provider.
- Governments that have not yet opened their market to competition may be hesitant to proceed with any market opening because of the instability of the industry.

- Governments that have allowed competition into their market may reevaluate the structure of their markets. In this regard, they may choose to limit the number of entrants as a way to protect consumers from the financial instability of service providers, as opposed to allowing unfettered competition.

Can wireless service providers fare better?

There is a widely held belief that over time telecommunications service providers that primarily operate in the wireless arena will fare better than traditional wireline service providers. However, for this belief to be validated, it is important to understand the rationales behind it. Traditionally wireless service providers have not had to absorb the large up-front costs associated with network builds that wireline providers do (with the exception of satellite service providers). Unfortunately, in today's environment, that is not always in the case. Many wireless providers, especially for mobile telephony, have committed to wide buildouts as an integral part of their concessions or regulatory authorizations. In order to ensure that the service provider remains viable, it may be necessary for the governments that obtained such commitments to revisit and revise them in light of the latest financial developments. Failure of governments to recognize economic realities may directly impact their ability to ensure their populations receive the sorts of services they had envisioned. For example, governments may find a way to provide financial credits to service providers that agree to provide service to rural or remote populations.

Further, with the advent of the auction of spectrum in many countries, wireless service providers have had to spend millions of dollars just to obtain a regulatory authorization to operate their system. This is on top of the monies that must be expended to build the wireless services network. It is important for governments to evaluate the impact of the use of spectrum auctions on the ability of service providers to offer their products on a cost-effective basis. The costs of the auction clearly impact the viability of the service. Therefore, governments that rely on auctions to award spectrum authorizations may want to examine other licensing processes and spectrum-management structures in order to encourage the deployment of new services. Such procedures may include the increased use of secondary markets and gray-spectrum markets [4].

In addition, many experts, especially in Europe, believe that the adoption of global standards on technology is necessary if cost-effective wireless services are going to be deployed. A good example of this is the past efforts of CEPT and its member countries to obtain a common frequency band for 3G services at WRC 2000. This position was advocated because many governments and companies believe that for deployment of a mass consumer spectrum-based service to be successful, large economies of scale must be captured in the development and manufacture of equipment. Having uniform technical characteristics (including frequency bands in which to operate) is a desirable attribute.

This approach, while having the benefit of creating a large global market for products, has many drawbacks. First, it will likely limit innovation. If companies are mandated to work with a single standard, it will be near impossible for companies that develop new and novel approaches to make it through the standard-setting process. Second, this approach calls into question the individual sovereignty of nations in making their own decisions concerning the use of the radiocommunications spectrum resource. Accordingly, many countries may be hesitant in allocating spectrum to a specific use, as opposed to a specific service.

Further, for the wireless industry and companies that rely on radiocommunications spectrum to emerge successfully from the current financial crunch, they need to have the initiative to figure out how to more efficiently use the radiocommunications spectrum resource. Increasingly, the problem is that demand for spectrum far outstrips the amount of available spectrum. Accordingly, one of the major challenges facing the spectrum industry is how to get more use out of spectrum that lies fallow a substantial amount of the time. The answer relies on the empowerment of technology that will allow for more innovative uses of existing technologies, like software-defined radio and spectrum sharing. In addition, governments should look at the continued use of unlicensed bands as a source of innovation for showing us the vision of alternative ways spectrum can be used that are outside the traditional model of spectrum-based telecommunications services.

It is also important that companies understand the unique needs of the population of consumers that they are serving. For example, the use of wireless services, such as mobile telephones, is much greater in countries such as Finland and Japan than in the United States. In this regard, wireless service providers and other telecommunications providers that rely upon the spectrum resource in their business need to deploy services that are responsive to the market.

Finally, it all comes down to one thing for success to take hold—profitability. This means that companies must accurately depict the economics associated with their business plan—including taking into account the lead time necessary for obtaining sufficient customers to begin to recognize a profit and realize adequate cash flow. The reason so many dot-coms went bankrupt in the early 2000s was that they were operating without a profit for an extended period of time and without sufficient cash flow. Experience has told us that unusually long lead times to profitability and operating without sufficient cash flow will not lead to a successful operation. Accordingly, it is important that companies that rely on the use of the radiocommunications spectrum develop sound business plans that are executed on a timely basis and in a cost-effective manner and that are responsive to consumer needs.

Endnotes

[1] See Bott, Jennifer, "Telecommunications: WorldCom's Collapse Mirror's Industry's," *Detroit Free Press,* July 1, 2002, available at http://www.freep.com/money/business/telcom1_20020701.htm.

[2] Rosenbush, Steve, "Telecommunications," *Business Week On-Line,* January 8, 2001, available at http://www.businessweek.com/2001/01_02/b3714089.htm.

[3] UMTS Forum, "Bright Future Predicted for UMTS/3G," available at http://www.umts-forum.org/mobilennium_online/2001.05/story1.htm. ("This [the UMTS Forum] study shows that 3G has a very bright future. This is important as it reaffirms our conviction that current concerns over 3G should not overshadow the very positive long term prospects for the technology," noted Paola Tonelli of the UMTS Forum. "The study represents the consensus view of over 200 companies within the industry. It is based on a very rigorous methodology and conservative assumptions.")

[4] The gray spectrum or *open* spectrum movement would be somewhat dependent on the development of *smart* transmitters that could allow the avoidance of interference between services, according to Gomes, Lee, "Visionaries See a Day When Radio Spectrum Isn't Scarce Commodity," *Wall Street Journal,* September 30, 2002, p. B1.

Appendix A
List of Web Addresses

ABC
http://www.abc.com

American National Standard Institute
http://www.ansi.org

Alcatel
http://www.alcatel.com

Anatel (Brazil)
http://www.anatel.gov.br

Asia Pacific Telecommunity
http://www.aptsec.org

AT&T
http://www.att.com

AT&T Wireless
http://www.attws.com

Australian Department of Foreign Affairs and Trade
http://www.dfat.gov.au

BBC
http://www.bbc.co.uk

BellSouth
http://www.bellsouth.com

British Telecommunications
http://www.bt.com

CBS
http://www.cbs.com

Cellular Telephone and Internet Association
http://www.ctia.org

Citibank
http://www.citibank.org

COFETEL (Mexico)
http://www.cft.gob.mx

Hughes
http://www.hughes.com

CITEL
http://www.citel.org

International Telecommunication Union
http://www.itu.int

Iridium
http://www.iridium.com

Korea Telecom
http://www.kt.co.kr

Motorola
http://www.motorola.com

NASA
http://www.nasa.gov

NTIA
http://www.ntia.doc.gov

NBC
http://www.nbc.com

Nokia
http://www.nokia.com

Nortel Networks
http://www.nortelnetworks.com

NATO
http://www.nato.int

PanAmSat
http://www.panamsat.com

Qualcomm
http://www.qualcomm.com

Samsung
http://www.samsung.com

Skystation
http://www.skystation.com

TIA
http://http://www.tiaonline.org

Teledesic Corporation
http://www.teledesic.com

Telefonica de Espana
http://www.telefonica.es

United Airlines
http://www.ual.com

U.K. Department of Trade and Industry
http://www.dti.gov.uk

U.S. Coast Guard
http://www.uscg.mil

U.S. Department of Commerce
http://www.doc.gov

U.S. Department of Defense (DOD)
http://www.defenselink.mil

U.S. Department of State, Bureau of Economic and Business Affairs, Communications, and Information
http://www.state.gov/e/eb/cip

U.S. Federal Aviation Administration (FAA)
http://www.faa.gov

U.S. Federal Communications Commission (FCC)
http://www.fcc.gov

Wireless Communications Association International
http://www.wcai.org

World Trade Organization (WTO)
http://www.wto.org

About the Author

Jennifer A. Manner is currently the director for international alliances in the Global Carrier Relations Group at WorldCom, Inc., where she is responsible for developing, implementing, and managing global alliances for the second-largest telecommunications long-distance service provider. Prior to this position, Ms. Manner was the associate counsel for foreign market access and international wireless services at WorldCom. In this position, Ms. Manner had responsibility for crafting the regulatory strategy for foreign market access for the provision of telecommunications services. In addition, Ms. Manner was responsible for the regulatory issues governing international wireless issues, including those at the International Telecommunication Union.

Prior to joining WorldCom, Ms. Manner was an associate in the communications group of Akin, Gump, Strauss, Hauer, and Feld, L.L.P., where she represented wireless (including satellite) and wireline telecommunications service providers, equipment manufacturers, and new technology service providers in international and domestic forums. Representative clients included Teledesic Corporation, Vebacom, Soros Funds, and AT&T Wireless Services. Before holding this position, Ms. Manner was an attorney-advisor at the Federal Communications Commission.

Ms. Manner is also currently an adjunct professor of law at Georgetown University Law Center, where she teaches international telecommunications regulation. Ms. Manner is also the chair of the Section on Computer and Telecommunications Law of the D.C. Bar Association and cochair of the Communications Law Forum of the D.C. Women's Bar Association. Ms. Manner is a member of the International Space Law Society and was formerly on the board of the U.S. ITU Association.

Ms. Manner has published and spoken extensively on the area of telecommunications regulation. Her first book, *Global Telecommunications Market Access*, was published by Artech House in 2002.

Ms. Manner holds an LL.M. in international law *with distinction* from Georgetown University Law Center, a J.D. *cum laude* from New York Law School, and a B.A. in political science and theatre arts from the State University of New York at Albany.

Ms. Manner lives in Kensington, Maryland, and Kent, Ohio, with her husband, Dr. Eric Glasgow, a professor at the Northeastern Ohio Universities College of Medicine.

Index

3G devices, 25–26
3G mobile services, 3, 25–29
  allocations, 27
  Europe licensing mechanisms, 118
  ITU study groups, 26
  spectrum requirements, 37, 46–47
  U.K. auction results, 4

Access, 12
  costs, 68
  demands, 134
  technical limitations to, 135
Adjunct uses, 113
Aeronautical fixed service (AFS) allocation, 41
Aeronautical mobile satellite service (AMSS) allocation, 41
Aeronautical mobile service (AMS) allocation, 41
Allocation
  3G, 27
  AFS, 41
  AMS, 41
  AMSS, 41
  defined, 9
  domestic, 72, 93
  EES, 42
  FS, 41
  FSS, 41
  international, 72, 83–89
  MS, 41
  MSS, 41, 113–14
  NGSO FSS, 25
  primary, 83
  scheme, 40–42
  secondary, 83–84
  space research service, 42
  Table of Frequency Allocations, 15–16, 39
  types, 83–84
Allotments
  defined, 9
  plans, 39
ANATEL, 106, 108
  formation, 108

ANATEL (*continued*)
mandate, 108
Asia-Pacific Telecommunity (APT), 77, 81–82
Assignment
auctions, 122–27
beauty contests, 127
comparative hearings, 119
defined, 10
domestic, 94
efficient use and, 116
financial viability and, 116
guides, 116–17
harmful interference avoidance and, 117
lotteries, 121–22
negotiated solutions, 119–21
plans, 39
processes overview, 118–27
public interest and, 117
to specific users, 116–27
technical viability and, 116
Auctions, 122–27
advantages/disadvantages of, 123–26
applicants, 122
efficient/high-value use and, 124–25
exception in U.S., 126–27
governments and, 118
international law and, 125
monetary, 118, 123
participation criteria, 122
public interest and, 124
revenue and, 125–26
speed, 123–24
transparency, 124
See also Assignment

Bandwidth, 36–37
defined, 36
example, 36–37
Beauty contests, 127
Broadcasters, 60
defined, 57
national, 60

Cellularvision, 22
Cofrequency sharing, 132–33, 143–46
defined, 143
as first compromise choice, 143
overview, 143–44
solutions, difficulty in reaching, 144–46
successful, 145
technical/operational rules and, 145
Commercial interest, 17–18
Comparative hearings, 119
Conference of European Posts and Telecommunications Administration (CEPT), 76, 77, 81
defined, 76, 81
member states, 81, 82
Conference preparatory meetings (CPMs), 88
Conflicts
minimizing potential for, 135–40
potential for harmful interference, 133–35
Consumers, 62–63
general, 62, 63
high-end, 62–63
as participants in international arena, 97–98
telecommunications industry and, 169
Costs
secondary markets and, 154
spectrum access, 68
unlicensed spectrum use and, 138
Customer migrations, 166

Department of Trade and Industry (DTI), 106
Designation
defined, 10–11
of spectrum to specified users, 112
Domestic allocation, 72, 93
of frequency bands to individual services, 110–16
regulation and, 93 *See also* Allocation
Domestic regulation, 93–129
cornerstones, 107–9

influences, 106
international representation, 93
overview, 93–94, 105–7
participation in international arena, 94–103
responsibilities, 93–94
See also Regulation

Earth exploration service (EES) allocation, 42
Efficiency, 116, 135
in assignment/licensing methods, 116, 137
secondary markets and, 153
standards role in, 137
technology, 137
unlicensed spectrum use and, 140

Federal Communications Commission (FCC), 18, 24, 113, 115
Flexible use, 113–14, 137
secondary markets and, 153, 159
unlicensed spectrum and, 138–39
Frequencies, 34–35
Frequency bands, 36
access, cost and delay, 43
cost differentials, 45
current allocation/uses of, 43–44
domestic allocation of, 110–16
not subjected to secondary markets, 160
propagation issues, 44
proposed, 42–43
underutilized, 136
use determination, 94
use of, 37
Frequency-band segmentation, 146–47
defined, 143
disadvantages, 146
example, 146–47
overview, 143–44
plans, 146
Frequency sharing
cofrequency, 132–33, 143–46
defined, 39
solutions, 144
unlicensed spectrum use and, 139
FS
allocation, 41
recognition, 85
FSS allocation, 41

Government
formal response review, 110
incentives, 135–36
interest, 18–19
monetary auctions and, 118
operational/technical requirements, 136–37
participants, 95–96
regulator, 64–66
secondary markets and, 154
subsidiary regulation goals, 75
telecommunications industry and, 169–70
use, 55–57
as user, 54
Gray spectrum, 4, 11, 137

Harmful interference, 37–40
assignment and, 117
avoidance, 38–39, 117
defined, 38
secondary markets and, 160
solutions, 131–49
See also Interference
High-altitude platform service (HAPS), 111

Identification
defined, 10
ITU Radio Regulations and, 26
of spectrum to specified users, 112
IMT 2000, 26, 27
Incentives, 135–36
new technology development, 136
underutilized frequency band, 136

Incumbent users, 17, 59
Inter-American Telecommunications Conference (CITEL), 77, 81
- defined, 77
- U.S. support and, 82–83

Interests, 17–20
- commercial, 17–18
- government, 18–19
- military, 19–20
- public, 117, 124

Interference
- amount, 38
- defined, 38
- electrical, 45
- harmful, 37–40, 117, 131–49, 160
- unlicensed spectrum use, 140
- wireless networks and, 53
- wireline networks and, 53

International allocation
- process, 83–89
- regions, 84–85
- technical issues, 85–86
- types, 83–84
- WRC process review, 87–89
- *See also* Allocation

International Frequency Registration Board (IFRB), 14
International meetings
- delegation members, 102–3
- delegation to, 101–3
- domestic preparatory process for, 98–103
- private-sector participation at, 103

International process
- consumers and, 97–98
- domestic participants in, 95–98
- governmental participants, 95–96
- private sector participants, 96–97

International Radiotelegraph Conference (IRC), 13
International Table of Frequency Allocations, 15–16
- footnotes, 41
- success, 15
- treaty status, 15
- *See also* Allocation

International Telecommunication Union (ITU), 5
- Constitution, 78
- Convention, 6
- Council, 83, 84
- defined, 5
- formation, 14
- member states, 5, 77
- organizational structure, 14
- overview, 78–80
- Radiocommunications Sector. *See* ITU-R
- Radio Regulations, 26, 27, 37, 38, 78, 86
- regulations, 6
- sectors, 78
- spectrum allocation, 5
- Table of Frequency Allocations, 86

Iridium, 66–67
ITU-R, 14, 15
- defined, 78
- illustrated, 80
- overview, 79–80
- procedures, 78
- Radiocommunications Bureau, 40, 79
- responsibilities, 79
- RRB, 79
- study groups, 79–80, 88
- *See also* International Telecommunication Union (ITU)

Local multipoint distribution service (LMDS), 22
Lotteries, 121–22

Market(s)
- demand, 66–67
- secondary, 151–61
- secondary, regulation of, 94, 128–29

Master International Register of Frequency Bands, 40
Military
- interest, 19–20

private industry conflict, 57
use, 56
MS
allocation, 41
recognition, 85
MSS allocation
defined, 41
NGSO, 113–14
Multipoint multichannel distribution systems (MMDS), 27, 28, 135

Negotiated solutions, 119–21
formal, 120
information, 119–20
processes, 119
relocation in, 115
success, 120–21
unsuccessful, 121
See also Assignment
Negotiations
formal, 142
informal, 142
New Zealand secondary market, 156–58
drawbacks, 157–58
as first secondary market country, 157
property rights, 157
spectrum manager, 157
See also Secondary markets
Nongeostationary orbit fixed satellite service (NGSO FSS), 3, 21–25, 114
advocates, 22–23
defined, 22
global designation of spectrum for, 23
Ka band for, 23
provisional global designation, 25
spectrum allocation, 25
Nongeostationary orbit (NGSO)
defined, 21
MSS, 22
satellite systems, 21
NTIA process, 99

OFTEL, 106

Personal communications service (PCS), 26, 114
Planned users, 17, 59
Primary allocation, 83
Private-sector coordinators, 141
Private-sector participants, 96–97, 102–3
in international arena, 96–97
at international meetings, 102–3
on national delegations, 102
as private-sector members, 103
Propagation, 44, 52
Public interest, 117, 124
secondary markets and, 154
standard, 117
Public safety uses, 21

Radio astronomers example, 20
Radiocommunications
issues, handling, 106
service categories, 111
Radiocommunications Bureau
harmful interference and, 40
responsibilities, 79
Radiocommunications spectrum. *See* Spectrum
Radio Regulation Board (RRB), 79
Refarming. *See* Relocation
Regulation, 71–90
domestic, 93–129
goals, 73, 74–76
goals, balancing, 75–76
influences, 106
reasons for, 71–74
regional organization impact on, 81–83
of secondary markets, 94, 128–29
subsidiary goals, 74–75
Regulatory bodies, 76–80
APT, 77, 81–82
CEPT, 76, 77, 81
CITEL, 77, 81
ITU-R, 78–80
Regulatory fees, 159
Regulatory process
domestic, 72, 73

Regulatory process (*continued*)
  international, 72, 73, 77–78
  time for, 128
Regulatory regime, 64–66, 71–90
Relocation, 114–15, 147–49
  compensation, 148–49
  controversy, 149
  defined, 147
  example, 116
  issues, 148–49
  as negotiated solution, 115
  reasons for, 114–15, 147
  regulators and, 115
  See also Users
Rulemaking, 142

Scarcity, spectrum, 2, 37–40, 133
Scientific community example, 20
Secondary allocation, 83–84
Secondary markets, 151–61
  advantages, 153
  advocates of, 152
  auctioned spectrum, 160
  commercial disputes, 158–59
  construction milestones, 159
  drawbacks, 153–54
  flexible use, 159
  frequency bands not subjected to, 160
  governments and, 154
  harmonizing operational rules, 160
  impact, 155
  increasing use of, 151–52
  licensee responsibility, 158
  New Zealand, 156–58
  protection from harmful
    interference, 160
  regime, creating, 158–61
  regime structure, 155–56
  regime types, 155–58
  regulation of, 94, 128–29
  regulatory fees, 159
  rules/regulations, 159
  secondary user responsibilities, 158
  use, limiting time/scope of, 158
Self-regulation scenarios, 54–55
Signal strength, for reliable service, 44
Spectrum
  3G requirements, 37, 46–47
  access, 2, 68, 134
  allocation, 9, 41, 72, 83–89, 93
  allotments, 9, 39
  assignment, 10
  availability, 4, 67
  battles, 3, 21–29
  caps on ownership, 160
  as commodity, 2–3
  comparable, need for, 148–49
  conflict, 133–40
  defined, 1
  designation of, 10–11, 112
  domestic issues, 6–8
  economics and, 134
  global coordination, 71–72
  gray, 4, 11, 137
  identification of, 10, 26, 112
  international overview, 5–6
  issues, 3
  management, 11–16
  overview, 4–8
  practical limits of, 45
  primer, 33–47
  regulation, 71–90, 93–129
  as resource, 34
  scarcity, 2, 37–40, 133
  technical characteristics, 34–37
  terminology, 8–11
  unlicensed usage, 138–40
  U.S. chart, 35
  use, 9–10, 42–46
  utilization, 2
  value, 4
  war, rationale, 33
Spectrum manager, 157

Table of Frequency Allocations
  allocations provided by, 39
  conformity to, 39
  domestic, 16
  international, 15–16, 41
  See also Allocation

Technical characteristics, 34–37
bandwidth, 36–37
frequencies, 35–36
frequency bands, 36, 37
Technical rules
changes to, 112–13
enforcement of, 94, 128
implementation of, 94, 127–28
Telecommunications equipment manufacturers, 60–62
activities, 61
activity motives, 61
constraints, 62
defined, 60
examples, 60–61
Telecommunications financial crisis, 163–72
business plan rethinking and, 164
impact of, 166–68
key reasons for, 165–66
overview, 163–65
rebuilding and, 168–72
Telecommunications industry
accounting practices, 165
bankrupt assets sale, 168
competition decrease, 168
consumers and, 169
customer migration, 166
debt-free competitors, 167–68
decline, xv
direction change, xvi–xvii
equipment development delay, 166
faulty forecast assumptions, 166
funding, 169
"givens," xvi
governments and, 169–70
new technology development, 169
outsourcing, 167
over buildout of capacity, 165
as participant in international arena, 96–97
quality of service decrease, 166–67
rebuilding, 168–72
reshaping itself, 164
resources for expansion, 169
retrenchment period, 169
tarnished reputation, 168
vendor payment, 167
Telecommunications service providers, 58–60
defined, 57
market aggression, 58
success, 59–60
Teledesic, 22–25, 59
Transmitters/receivers, operating characteristics, 45

Unilateral actions, 142
Universal Mobile Telecommunications System (UMTS), 28
Unlicensed spectrum use, 138–40
advantages, 138–39
cost savings, 138
drawbacks, 139–40
flexibility, 138–39
inefficiency, 140
innovation, 139
interference potential, 140
mobility, 139
overuse potential, 140
spectrum preservation, 139
spectrum sharing, 139
unpredictability, 139
U.S.
exception to auctions in, 126–27
Office of Communications and Information Policy (CIP), 100
regulation goals, balancing, 75–76
spectrum chart, 35
Users
assignment to, 116–27
designation of spectrum to, 112
domestic government, 54
high-end, 62–63
identification of spectrum to, 112
importance/priority, 136
incumbent, 17, 59
planned, 17, 59
relocation of, 114–16, 147–49
spectrum value to, 67

Use(s)
adjunct, 113
amount of spectrum required for, 43
current, 43–44
defined, 9–10
efficient, 116, 135
evaluation considerations, 42–46
flexible, 113–14, 137
government, 55–57
impact factors, 63–69
military, 56
public safety, 21
unlicensed spectrum, 138–40

Very small aperture terminal (VSAT), 12

Wideband code division multiple access (WCDMA), 29
Wireless networks, 49–54
advantages, 50–52
anticompetitive concerns, 51–52
anytime, anywhere communications, 51
costs, 50
deployment, 50–51
disadvantages, 52–53
geographic reach, 50
infrastructure availability, 68–69
interference, 52
mobility, 50
propagation, 52
regulation, 52
regulatory fees, 52–53
roof rights, 52
wireline network reliance, 53
Wireless telecommunications services, 1
Wireline networks
advantages, 53
disadvantages, 52–53
wireless network reliance on, 53
World Radiocommunications Conferences (WRCs), 14
agendas, 84, 87
attendance, 87
committees, 88
coordinating positions of, 82
meetings, 77
member states, 87
process, 78, 90
process overview, 87–89
responsibilities, 110
success, 89
WRC 92, 86
WRC 95, 24, 25
WRC 97, 25, 86
WRC 2000, 26, 28, 29, 65, 80, 100–101
World Trade Organization (WTO)
commitments, 141
markets, 109
World Trade Organization (WTO) Agreement, 65, 107
coverage, 108
impact of, 108–9
implementation areas, 109
signatories, 108

Recent Titles in the Artech House Telecommunications Library

Vinton G. Cerf, Senior Series Editor

Access Networks: Technology and V5 Interfacing, Alex Gillespie

Achieving Global Information Networking, Eve L. Varma et al.

Advanced High-Frequency Radio Communications, Eric E. Johnson et al.

ATM Interworking in Broadband Wireless Applications, M. Sreetharan and S. Subramaniam

ATM Switches, Edwin R. Coover

ATM Switching Systems, Thomas M. Chen and Stephen S. Liu

Broadband Access Technology, Interfaces, and Management, Alex Gillespie

Broadband Networking: ATM, SDH, and SONET, Mike Sexton and Andy Reid

Broadband Telecommunications Technology, Second Edition, Byeong Lee, Minho Kang, and Jonghee Lee

The Business Case for Web-Based Training, Tammy Whalen and David Wright

Centrex or PBX: The Impact of IP, John R. Abrahams and Mauro Lollo

Chinese Telecommunications Policy, Xu Yan and Douglas Pitt

Communication and Computing for Distributed Multimedia Systems, Guojun Lu

Communications Technology Guide for Business, Richard Downey, Seán Boland, and Phillip Walsh

Community Networks: Lessons from Blacksburg, Virginia, Second Edition, Andrew M. Cohill and Andrea Kavanaugh, editors

Component-Based Network System Engineering, Mark Norris, Rob Davis, and Alan Pengelly

Computer Telephony Integration, Second Edition, Rob Walters

Desktop Encyclopedia of the Internet, Nathan J. Muller

Digital Clocks for Synchronization and Communications, Masami Kihara, Sadayasu Ono, and Pekka Eskelinen

Digital Modulation Techniques, Fuqin Xiong

E-Commerce Systems Architecture and Applications, Wasim E. Rajput

Engineering Internet QoS, Sanjay Jha and Mahbub Hassan

Error-Control Block Codes for Communications Engineers, L. H. Charles Lee

FAX: Facsimile Technology and Systems, Third Edition, Kenneth R. McConnell, Dennis Bodson, and Stephen Urban

Fundamentals of Network Security, John E. Canavan

Gigabit Ethernet Technology and Applications, Mark Norris

Guide to ATM Systems and Technology, Mohammad A. Rahman

A Guide to the TCP/IP Protocol Suite, Floyd Wilder

Information Superhighways Revisited: The Economics of Multimedia, Bruce Egan

Integrated Broadband Networks: TCP/IP, ATM, SDH/SONET, and WDM/Optics, Byeong Gi Lee and Woojune Kim

Internet E-mail: Protocols, Standards, and Implementation, Lawrence Hughes

Introduction to Telecommunications Network Engineering, Tarmo Anttalainen

Introduction to Telephones and Telephone Systems, Third Edition, A. Michael Noll

An Introduction to U.S. Telecommunications Law, Second Edition Charles H. Kennedy

IP Convergence: The Next Revolution in Telecommunications, Nathan J. Muller

The Law and Regulation of Telecommunications Carriers, Henk Brands and Evan T. Leo

Managing Internet-Driven Change in International Telecommunications, Rob Frieden

Marketing Telecommunications Services: New Approaches for a Changing Environment, Karen G. Strouse

Multimedia Communications Networks: Technologies and Services, Mallikarjun Tatipamula and Bhumip Khashnabish, editors

Next Generation Intelligent Networks, Johan Zuidweg

Performance Evaluation of Communication Networks, Gary N. Higginbottom

Performance of TCP/IP over ATM Networks, Mahbub Hassan and Mohammed Atiquzzaman

Practical Guide for Implementing Secure Intranets and Extranets, Kaustubh M. Phaltankar

Practical Internet Law for Business, Kurt M. Saunders

Practical Multiservice LANs: ATM and RF Broadband, Ernest O. Tunmann

Principles of Modern Communications Technology, A. Michael Noll

Protocol Management in Computer Networking, Philippe Byrnes

Pulse Code Modulation Systems Design, William N. Waggener

Security, Rights, and Liabilities in E-Commerce, Jeffrey H. Matsuura

Service Level Management for Enterprise Networks, Lundy Lewis

SIP: Understanding the Session Initiation Protocol, Alan B. Johnston

Smart Card Security and Applications, Second Edition, Mike Hendry

SNMP-Based ATM Network Management, Heng Pan

Spectrum Wars:The Policy and Technology Debate, Jennifer A. Manner

Strategic Management in Telecommunications, James K. Shaw

Strategies for Success in the New Telecommunications Marketplace, Karen G. Strouse

Successful Business Strategies Using Telecommunications Services, Martin F. Bartholomew

Telecommunications Cost Management, S. C. Strother

Telecommunications Department Management, Robert A. Gable

Telecommunications Deregulation and the Information Economy, Second Edition, James K. Shaw

Telemetry Systems Engineering, Frank Carden, Russell Jedlicka, and Robert Henry

Telephone Switching Systems, Richard A. Thompson

Understanding Modern Telecommunications and the Information Superhighway, John G. Nellist and Elliott M. Gilbert

Understanding Networking Technology: Concepts, Terms, and Trends, Second Edition, Mark Norris

Videoconferencing and Videotelephony: Technology and Standards, Second Edition, Richard Schaphorst

Visual Telephony, Edward A. Daly and Kathleen J. Hansell

Wide-Area Data Network Performance Engineering, Robert G. Cole and Ravi Ramaswamy

Winning Telco Customers Using Marketing Databases, Rob Mattison

World-Class Telecommunications Service Development, Ellen P. Ward

For further information on these and other Artech House titles, including previously considered out-of-print books now available through our In-Print-Forever® (IPF®) program, contact:

Artech House
685 Canton Street
Norwood, MA 02062
Phone: 781-769-9750
Fax: 781-769-6334
e-mail: artech@artechhouse.com

Artech House
46 Gillingham Street
London SW1V 1AH UK
Phone: +44 (0)20 7596-8750
Fax: +44 (0)20 7630-0166
e-mail: artech-uk@artechhouse.com

Find us on the World Wide Web at:
www.artechhouse.com